NATIONAL DEFENSE RESEARCH INSTITUTE

A Strategic Assessment of the Future of U.S. Navy Ship Maintenance

Challenges and Opportunities

Bradley Martin, Michael E. McMahon, Jessie Riposo, James G. Kallimani, Angelena Bohman, Alyssa Ramos, Abby Schendt

Prepared for the United States Navy
Approved for public release; distribution unlimited

For more information on this publication, visit www.rand.org/t/RR1951

Library of Congress Cataloging-in-Publication Data is available for this publication.
ISBN: 978-0-8330-9923-5

Published by the RAND Corporation, Santa Monica, Calif.
© Copyright 2017 RAND Corporation
RAND® is a registered trademark.

*Cover top: photos by U.S. Navy;
bottom: photo courtesy of Huntington Ingalls.*

Limited Print and Electronic Distribution Rights

This document and trademark(s) contained herein are protected by law. This representation of RAND intellectual property is provided for noncommercial use only. Unauthorized posting of this publication online is prohibited. Permission is given to duplicate this document for personal use only, as long as it is unaltered and complete. Permission is required from RAND to reproduce, or reuse in another form, any of its research documents for commercial use. For information on reprint and linking permissions, please visit www.rand.org/pubs/permissions.

The RAND Corporation is a research organization that develops solutions to public policy challenges to help make communities throughout the world safer and more secure, healthier and more prosperous. RAND is nonprofit, nonpartisan, and committed to the public interest.

RAND's publications do not necessarily reflect the opinions of its research clients and sponsors.

Support RAND
Make a tax-deductible charitable contribution at
www.rand.org/giving/contribute

www.rand.org

Preface

The U.S. Navy's ship inventory and the shipbuilding and repair industrial base that supports these ships have experienced significant changes over the previous three decades. In the next 30 years, significant changes to the fleet composition and the maintenance requirements of the fleet are likely to occur. For example, there will likely be an increased number of littoral combat ships. These have distinctly different maintenance requirements from other platforms in the fleet.

However, as the fleet has declined in the past, so has the number of maintenance providers. In 1993, the Navy had eight public shipyards. Today there are four. These naval shipyards are almost exclusively focused on supporting nuclear-powered aircraft carriers and submarines. The work conducted at the public shipyards that once also maintained surface ships has largely transitioned to the private-sector providers.

To ensure that the private-sector industrial base is available and able to support the Navy's future maintenance and modernization requirements, the Navy must understand the future maintenance needs and develop a strategic approach to ensure that the necessary capabilities—including facilities, engineers, and trade labor—are available. RAND Corporation researchers assisted the Commander of the Naval Sea Systems Command (NAVSEA) to assess possible supply and demand capabilities in the ship maintenance workload and note long-term challenges facing mitigation efforts. This report also offers a number of recommendations for NAVSEA and Department of Defense leadership to consider to mitigate challenges and plan strategically for the coming years.

This research was sponsored by the Commander of NAVSEA and conducted within the Acquisition and Technology Policy Center of the RAND National Defense Research Institute, a federally funded research and development center sponsored by the Office of the Secretary of Defense, the Joint Staff, the Unified Combatant Commands, the Navy, the Marine Corps, the defense agencies, and the defense Intelligence Community.

For more information on the RAND Acquisition and Technology Policy Center, see www.rand.org/nsrd/ndri/centers/atp or contact the director (contact information is provided on the web page).

Contents

Preface iii
Figures and Tables vii
Summary ix
Acknowledgments xvii
Abbreviations xix

CHAPTER ONE
Introduction 1
Approach and Methodology 4
Study Limitations 6
Organization of This Report 8

CHAPTER TWO
The Future Demand for U.S. Navy Ship Repair 9
The Elements of Demand 9

CHAPTER THREE
The U.S. Navy Ship Maintenance Industrial Base 33
Defining the Maintenance Industrial Base 34
Industrial Base Assessment 39
Potential Vulnerabilities 45
Summary 46

CHAPTER FOUR
Industry Incentives Within the Maintenance Industrial Base 49
Planning Horizons 50

Contracting Mechanisms 51
Labor Market and Infrastructure Challenges................................. 52
Market Structure 55
Summary 56

CHAPTER FIVE
Demand-Supply Mismatches 59
Labor Market Shortfalls 59
Facilities 61
Summary 66

CHAPTER SIX
Conclusions and Recommendations 67

APPENDIXES
**A. Shipbuilding and Maintenance Capabilities in the United
States, by Region and Shipyard** 73
B. Cleaning VAMOSC Data 89

References 91

Figures and Tables

Figures

1.1.	Total Number of Active Ships in the U.S. Navy from 1993 to 2016	2
1.2.	Mismatch Between Budget and Programming	7
2.1.	Future Naval Ship Force Structure	13
2.2.	DDG-51 Work Distribution in the TFP	15
2.3.	SWLIN Distribution in Representative DDG-51 Availabilities	16
2.4.	*Virginia*-Class (SSN-774) Maintenance Plan	17
2.5.	DDG-51 Performed Versus TFP-Directed Maintenance	18
2.6.	CG-47 Performed Versus CMP-Directed Maintenance	19
2.7.	Difference Between CMP and Actual Budgeted Man-Days	21
2.8.	Difference Between CMPs and Programmed Levels in Private- and Public-Shipyard Availabilities	22
2.9.	Planned Maintenance Workloads	23
2.10.	Private-Sector Equipment Distribution	25
2.11.	Added Maintenance with Regular Recovery of Deferred Maintenance	27
2.12.	Added Maintenance with Five-Year Deferral of Maintenance Backlogs	28
2.13.	Added Maintenance with Ten-Year Deferral of Maintenance Backlogs	29
3.1.	Public Shipyard and Support Locations	35
3.2.	Shipyards Supporting U.S. Navy Warships	38
4.1.	Estimate of Current Range of Mayport Ship Repair Capacity in Man-Days	54
4.2.	Consolidation of Ship Repair Providers into GD NASSCO	55

5.1.	East Coast Dry-Dock Demand	64
5.2.	West Coast Dry-Dock Demand	65
A.1.	PSNS & IMF	74
A.2.	PHNS & IMF	79
A.3.	PNSY	81
A.4.	NNSY	83

Tables

2.1.	Annual New Construction Plan	11
3.1.	Civilian End Strength and Estimated Employment Levels at Public and Private Shipyards, by Region, in 2016	40
3.2.	Bureau of Labor Statistics Outlook on Ship Repair Trades, 2014–2024	41
3.3.	Man-Days Required per Year in the Port to Maintain the Private-Sector Capabilities-Estimated by Ship Repair Associations	42
3.4.	Number of Docks That Can Accommodate Each Ship Class, by Region, as of 2016	44
3.5.	FaC Assessment Methodology Example	45

Summary

The U.S. Navy's ship inventory and the shipbuilding and repair industrial base that supports these ships have experienced significant changes over the previous three decades. The number of ships in the fleet declined, from a total 454 active ships in 1993 to a low of 271 in 2015. However, the Navy's most recent Long-Range Shipbuilding Plan suggests that changes to the fleet composition and the maintenance requirements of the fleet are likely to occur in the next 30 years.[1] Specifically, there will be an increased number of littoral combat ships (LCSs), which have distinctly different maintenance requirements from other platforms in the fleet.

However, as the fleet has declined in the past, so has the number of maintenance providers. In 1993, the U.S. Navy had nine public shipyards. Today there are four. These naval shipyards are almost exclusively focused on supporting nuclear-powered aircraft carriers and submarines. The work conducted at the public shipyards that once also maintained surface ships has largely transitioned to the private-sector providers.

To ensure that the private-sector industrial base is available and able to support the Navy's future maintenance and modernization requirements, the Navy must understand the future maintenance needs and develop a strategic approach to ensure that the necessary capabilities—including facilities, engineers, and trade labor—are available. RAND Corporation researchers assisted the Commander

---

[1] Office of the Chief of Naval Operations, *Report to Congress on the Annual Long-Range Plan for Construction of Naval Vessels for Fiscal Year 2017*, Washington, D.C., July 2016.

of the Naval Sea Systems Command (NAVSEA) to reach these goals through three interrelated tasks: (1) estimate future workload demands, (2) characterize the current repair and modernization industrial base capacity, and (3) compare the supply and the demand of resources to identify potential misalignments.

Method and Research Approach

We used a number of approaches to answer the research questions. For the projection of demand, we used the Navy's 30-year shipbuilding plan and matched this to the class maintenance plans (CMPs) or the technical foundation papers (TFPs) for each of the ship classes. The CMP or the TFP is held to be the planned level of maintenance required for the lifetime of the force. Although CMPs provide only a schedule of maintenance availabilities, TFPs are more detailed and in fact break maintenance demand to the Ship's Work Line Item Number (SWLIN) level, which allows a more detailed understanding of demand components. SWLINs allow us to see not only the amount of work being done but also additional detail on the type of work, allowing us to understand future demand for shipyard capacity and labor.

Historically, there has been a propensity for the Navy to defer maintenance, with a resulting impact on both the current workload and the amount of work that needs to be executed across ship service lives; therefore, we used historical trends to project various scenarios of deferral to show the resulting impact. Using different assumptions regarding how the Navy elects to retire the backlog of deferred maintenance, we show how these scenarios result in different required maintenance levels over the near and long term.

To estimate future resources for supply, we first conducted a survey of the current capacity in public- and private-sector maintenance providers, looking at both key facility inventories and labor. For key facilities, we looked in detail at dry docks, matching availability to the expected CMP or TFP docking requirements. Although we could make some projections concerning expected future labor force from the Bureau of Labor Statistics and other sources to estimate capacity in public shipyards, the availability of labor and infrastructure from private-sector providers would depend primarily on choices made by private, for-profit companies. To gain insight into their decisions, we used interviews with industry management to better understand the incentives and disincentives for making investment decisions.

Demand for Maintenance Skilled Labor and Facilities Will Likely Increase

The Navy manages maintenance and modernization on all its ships throughout each ship's service life. The demand for maintenance services generated depends on several factors. The first factor is, simply enough, the force structure number and mix of platforms. The Navy will maintain some number of aircraft carriers, submarines, surface combatants, amphibious ships, and auxiliaries intended to satisfy presence and warfighting requirements. All are built with an expected service life; all will require maintenance and modernization throughout their service lives. The second component is what actually must be accomplished on these ships and submarines, tailored to each platform type, to reach service life. Documents such as the NAVSEA's TFPs and CMPs describe the work required at different stages of a ship's life, including dedicated maintenance periods requiring dry-docking and major modernizations.[2]

If the 30-year shipbuilding plan is executed and the Navy makes a consistent effort to comply with its CMPs, the long-range future maintenance workload will remain at least at current levels. Historical trends suggest that higher maintenance levels are likely. This projection applies in both public and private sectors.

The type of workload and, hence, the labor skills expected to be required are also not likely to change, with a similar distribution by

---

[2] NAVSEA, *Surface Maintenance Engineering Planning Program Class Depot Maintenance Technical Foundation Paper*, Washington, D.C., various years and for different classes; NAVSEA, *Surface Maintenance Engineering Planning Program Technical Foundation Paper, LCS1*, Washington, D.C., April 6, 2015a; NAVSEA, *Surface Maintenance Engineering Planning Program Technical Foundation Paper, LCS2*, Washington, D.C., May 4, 2015b.

SWLIN items appearing consistent in the decades to come. This indicates that trade-labor demand by skill will continue to require similar skills to current trades and also new skills to maintain fiber optics systems, photonics, control systems software, and power electronics.

Demands for facilities, in particular, dry docks, will be significant and, at times, overstress available dry docks by port, but the dry-dock demand might be met by allowing coast-wide bidding for dry-dock availabilities. The dry-dock demand predicted currently for the LCS-1 and LCS-2 classes of ships (littoral combat ships), when analyzed by homeport, does not appear executable within available facilities within homeport.

Deferral of maintenance actions will complicate management of maintenance demands. Deferrals occur for a variety of reasons, to include funding shortfalls, scheduling demands, and capacity shortfalls, and it is unrealistic to simply insist that they not occur. However, it is important to understand the impact. According to our historical data, the Navy has shown a tendency to defer maintenance on the two classes of surface ships examined. Analysis of the Future Years Defense Plan shows, conversely, that the Navy is planning on spending more than what the technical requirements would have dictated. Our models indicate that this is likely due to an attempt to recover lost maintenance and that the impact on out-year requirements becomes more severe the longer the maintenance is deferred.

Supply Analysis Shows That the Navy Primarily Influences Public Capabilities, with Less Influence on Private Ones

The U.S. Navy ship maintenance industrial base consists of a number of both public and private providers. The public sector is primarily focused on providing maintenance services to nuclear-powered ships, while the private sector is primarily focused on providing maintenance services to nonnuclear ships. There are two exceptions: Huntington Ingalls Industries–Newport News Shipbuilding performs refueling and complex overhaul of the nuclear aircraft carriers and General

Dynamics Electric Boat, which provides maintenance services to nuclear submarines.

The Department of the Navy has a large degree of control over the capabilities that are currently and will be provided by the public sector. The Navy determines the mission and function of the public support organizations. The Navy also establishes the composition and level of workforce required to accomplish the organizations' missions and identifies and makes the investments required to ensure that the necessary capabilities are provided. The Navy has less control over the capabilities that will be provided by the private sector. The private sector responds to market forces, which, in some cases, the Navy can influence.

Demand-Supply Mismatches Have Been Addressed in the Past, but Challenges Remain

While the Navy directly controls what happens in public shipyards in terms of workforce and infrastructure development, it relies on private industry to make plans and deliver services for surface ships and a large portion of the budget for aircraft carriers. The Navy cannot compel the delivery of these services; it has to create incentives for industry to not just deliver services but to make capital and personnel-development investments to meet needs over the long term.

This is not to say that industry concerns are necessarily congruent with Navy concerns. However, if the Navy would like industry to act in a particular way, it must find a way to convince industry that the reward for doing business with the Navy is sufficient to offset the concerns. The Navy has inaugurated several regimes over the years for managing its relationship with civilian providers. These have ranged from multi-ship, multi-year arrangements that promote teaming and long-term relationships between providers and the Navy to competition-based, firm fixed-price contracts intended to encourage competitive bidding. These management and policy choices have been accompanied by changes in planning timelines, changes in the organizations charged with oversight, and changes in the expected relationship between competitors. It is not clear that any of these have had sufficient time to actually be effective, and indeed one consistent observation is that the regimes change frequently.

Industry management repeatedly voiced concern over the impact of insufficient planning time for both short- and long-term decision-making. Industry has claimed that it generates the best product for the Navy when it receives sufficient time prior to availability start to provide a tailored and detailed work-package proposal. However, currently, the short timelines between request for proposal and need to begin the availability (contract award), combined with the uncertainty of the amount of future work, are particularly challenging. This uncertainty about future work also diminishes the incentive to make long-term capital investments, such as dry docks. Industry management expressed concern regarding the quality of the work-specification package that is being provided by the third-party industry team that is producing it for the regional maintenance center. The late contract award relative to the availability start date, in addition to a work-specification package that was likely to have changes to it upon execution, represented significant schedule risk to the Navy.

A primary concern presented by some repair associations and their contractor base was a lack of consideration of the industrial base and sustainment issues in the ship homeporting assignment process. An example is the assignment of an amphibious ready group that deploys together into a homeport, representing a major fraction of the maintenance workload in the harbor. This construct presents sharp workload-profile changes from overload conditions to workload levels below the minimum sustainable without large layoffs. Local contractors have had such cycles and observed that workers let go and not brought back within a few months never return and pursue other work paths. Our analysis suggests that repair and maintenance providers may face challenges in attracting sufficient numbers of qualified trade workers in the future, underlining the importance of careful planning for future demands.

Recommendations

Our recommendations to the Navy are the following:

- Work to establish a more integrated picture of port-wide maintenance demands.
- Identify work at public shipyards that is likely to be outsourced as early as possible in the planning cycle.
- Identify expectations for private-sector providers and create incentives for industry to support the plan.
- Explore public-private partnerships as a means to achieve cost and schedule goals.
- Develop partnered programs for developing ship repairs with specific skill bases.

The shipyard maintenance industrial base faces challenges in the future. There are limitations to how quickly the industrial base can grow before additional constraints and productivity barriers are reached. There is also a cost to sustaining an industrial base that is constantly going through boom and bust cycles. To improve decisionmaking capabilities, it is important for the Navy to develop an integrated picture of public- and private-sector workload, including commercial, U.S. Coast Guard, Military Sealift Command, and any other work in the port. Where the construction yards are relied on to assist with critical maintenance activities in times of need, the already-existing construction workload at the private shipyards should be considered. To achieve this end, it is critical to establish a good relationship with the private sector. While competition is desired, the number of providers in the space is limited, and without a significant commercial market, competition, or lack thereof, will be determined by the U.S. Navy. Although public-private partnerships can be difficult to implement and can exist in many forms, there are significant potential benefits to both the government and industry when implemented well, which the Navy should consider. For example, identifying investments in facilities and people to accommodate existing operational and maintenance schedules could become a more cooperative endeavor. The Navy could secure capacity, and industry could obtain more stability. There are many possible problems with public-private partnerships as well. This option would require additional and significant investigation to determine viability; however, early insight into this option suggests that such effort may be worth pursuing.

Acknowledgments

We would like to thank the civilian and military personnel of the Naval Sea Systems Command for their generous support of this study, as well as the four shipyards that supported our research-related visits: Norfolk Naval Shipyard, Puget Sound Naval Shipyard, Pearl Harbor Naval Shipyard, and Portsmouth Naval Shipyard. We would also like to thank Jacksonville Ship Repair Association; Puget Sound Ship Repair Association; BAE Systems; Oceaneering, Marine Hydraulics International, Huntington Ingalls–Newport News Shipbuilding; Southwest Regional Maintenance Center; Commander, Navy Regional Maintenance Center; Virginia Ship Repair Association; and Jeff Brooks. We also wish to thank Charles Goldman, Phil Pardue, and Brian Persons for their very insightful and helpful reviews. From RAND, we thank Danny Tremblay for his budget work and Sunny Bhatt and Christina Dozier for their administrative support.

Abbreviations

ABR	Agreement for Boat Repair
CG	guided-missile cruiser
CMP	class maintenance plan
CMSD	Continental Maritime of San Diego
CNRMC	Commander, Navy Regional Maintenance Center
CVN	nuclear-powered aircraft carrier
DDG	guided-missile destroyer
DSRA	Docking Selected Restricted Availability
EDSRA	Extended Docking Selected Restricted Availability
ESRA	Extended Selected Restricted Availability
FaC	Fragility and Criticality
FFG	frigate, guided missile
FY	fiscal year
FYDP	Future Years Defense Plan
GD NASSCO	General Dynamics NASSCO
HII-NNS	Huntington Ingalls Industries–Newport News Shipbuilding
LCS	littoral combat ship
LHA	landing helicopter assault (amphibious assault ship)

LHD	landing helicopter dock (amphibious assault ship)
LPD	amphibious transport dock (amphibious landing ship)
LSD	dock landing ship (amphibious landing ship)
MCM	mine countermeasures ship
MHI	Marine Hydraulics International
MSC	Military Sealift Command
MSMO	multi-ship multi-option (contract)
MSRA	Master Ship Repair Agreement
NAVSEA	Naval Sea Systems Command
NNSY	Norfolk Naval Shipyard
NSY	naval shipyard
OPNAV	Office of the Chief of Naval Operations
PB	President's Budget
PC	patrol ship
PHNS & IMF	Pearl Harbor Naval Shipyard and Intermediate Maintenance Facility
PSNS & IMF	Puget Sound Naval Shipyard and Intermediate Maintenance Facility
PNSY	Portsmouth Naval Shipyard
RCOH	refueling and complex overhaul
RMC	regional maintenance center
SRA	Selected Restricted Availability
SSBN	nuclear-powered ballistic missile submarine
SSN	nuclear-powered fast-attack submarine
SSGN	nuclear-powered cruise-missile submarine
SWLIN	Ship Work Line Item Number
TFP	technical foundation paper
VAMOSC	Visibility and Management of Operations and Support Costs

CHAPTER ONE

Introduction

The U.S. Navy's ship inventory has experienced significant changes over the previous three decades. The number of ships in the fleet declined, from a total 454 active ships in 1993 to a low of 271 in 2015, shown in Figure 1.1. During this period, the composition of the fleet also changed. The proportion of guided-missile destroyers (DDGs) in the fleet increased while the proportion of cruisers decreased. New classes of ships, such as the *Freedom*-class (LCS-1) and *Independence*-class (LCS-2) littoral combat ships, the *Virginia*-class (SSN-774) nuclear-powered fast-attack submarines, the *San Antonio*–class (LPD17) amphibious landing ships, and the *Ford*-class (CVN-78) nuclear powered-aircraft carriers were introduced. Other classes of ships, such as the cruise missile submarines, were introduced following conversion of the four oldest *Ohio*-class (SSBN-726) nuclear-powered ballistic missile submarines. Other classes, such as the *Los Angeles*–class (SSN-688) nuclear-powered fast-attack submarine, began retiring. The maintenance and modernization industrial base evolves to the numbers and types of platforms in the fleet at any time. Therefore, industrial base is linked to some degree to the Navy's Long-Range Shipbuilding (and decommissioning) Plan.

The shipbuilding and industrial base to support these ships has also changed. As the fleet has declined, so has the number of maintenance providers. In 1993, the U.S. Navy had eight public shipyards.[1] At the conclusion of the Cold War, the U.S. Department of Defense

---

[1] These were Puget Sound Naval Shipyard, Portsmouth Naval Shipyard, Norfolk Naval Shipyard, Pearl Harbor Naval Shipyard, Charleston Naval Shipyard, Mare Island Naval

Figure 1.1
Total Number of Active Ships in the U.S. Navy from 1993 to 2016

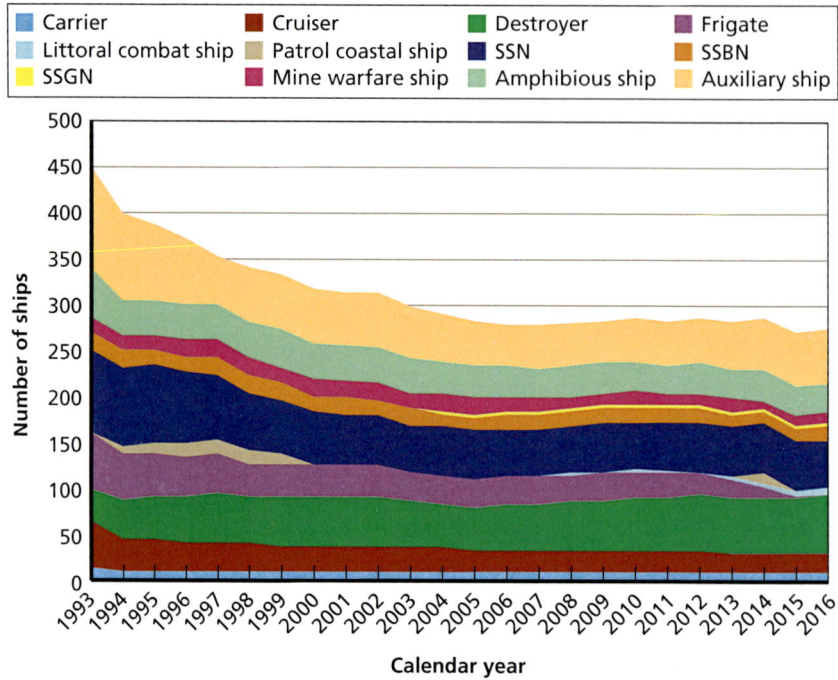

SOURCE: Annual expenditure calculations from the Navy Center for Cost Analysis's Navy Visibility and Management of Operating and Support (VAMOSC) database, fiscal years (FYs) 1993–2016.
RAND RR1951-1.1

determined that the existing capacity was in excess of need and thus proceeded to close several public shipyards. Closures focused primarily on those shipyards supporting nonnuclear platforms but included one shipyard that conducted nuclear submarine repair and overhaul. Today there are four public shipyards: Norfolk Naval Shipyard (NNSY), Portsmouth Naval Shipyard (PNSY), Pearl Harbor Naval Shipyard and Intermediate Maintenance Facility (PHNS & IMF), and Puget Sound Naval Shipyard and Intermediate Maintenance Facility (PSNS

Shipyard, Philadelphia Naval Shipyard, and Long Beach Naval Shipyard. See Shipbuilding History, "Public Shipyards," web page, undated.

& IMF). These naval shipyards (NSYs) are almost exclusively focused on supporting nuclear-powered aircraft carriers and submarines. The work conducted at the public shipyards that once also maintained surface ships has largely transitioned to the private-sector providers, including shipbuilders. While there is no single authoritative historical record of the number of ship repair and modernization providers in the private sector throughout this period, observations of the history of some of the current providers reveal extensive change in the provider base and consolidation within the industry.[2]

In the next 30 years, other significant changes to the fleet composition and the maintenance requirements of the fleet are likely to occur. These are detailed by the Navy's Long-Range Shipbuilding Plan.[3] As we discuss in Chapter Two, the demand for maintenance is expected to increase. There will be an increased number of LCSs, which have distinctly different maintenance requirements from other platforms in the fleet. There are now more-complex warships to maintain, such as the *Zumwalt*-class (DDG-1000) destroyers and the *Virginia*-class (SSN-774) nuclear submarines, as well as near-term periods where a large number of midlife availabilities for the *Arleigh Burke*–class (DDG-51) destroyers will occur. Unlike in the past, new classes of attack submarines and ballistic missile submarines will no longer require nuclear refueling. Total force-level inventories have declined over time; however, there are current proposals that may drive fleet-inventory increases

The Navy can, to some degree, control the capacity of the public shipyards by increasing the workforce to ensure that the necessary workload can be executed. The private sector, however, requires some confidence in a business-base workload and financial incentive to maintain and develop new capacity to meet the future repair and modernization needs of the Navy. To ensure that the private-sector industrial base is available and capable to support the Navy's future maintenance and modernization requirements, the Navy must assess the future maintenance needs and develop a more strategic approach to guaranteeing

---

[2] We discuss the consolidation of the industry in Chapter Three.

[3] Office of the Chief of Naval Operations, *Report to Congress on the Annual Long-Range Plan for Construction of Naval Vessels for Fiscal Year 2017*, Washington, D.C., July 2016.

that the necessary capabilities, including facilities, engineers, and trade labor, are available.

Approach and Methodology

The RAND National Defense Research Institute worked closely with the various Navy and private-sector organizations involved with the materiel support of ships in service. These included the Naval Sea Systems Command (NAVSEA), various program executive offices, U.S. Fleet Forces Command, NSYs, Navy maintenance and modernization managers, private-sector maintenance providers, and local and national ship repair associations.

To assist the Commander, NAVSEA, we pursued four primary tasks:

- **Estimate future workload demands:** For the projection of demand, we used the Navy's 30-year ship-building plan and matched this to the class maintenance plans (CMPs) or the technical foundation papers (TFPs) for each of the ship classes. This is held to be the planned level of maintenance required for the lifetime of the force. Although CMPs provide only a schedule of maintenance availabilities, TFPs are more detailed and break maintenance demand to the Ship's Work Line Item Number (SWLIN) level, which allows a more detailed understanding of demand components. SWLINs allow us to see not only the amount of work being done but also additional detail about the type of work, allowing us to understand future demand for shipyard capacity and labor.
    - Historically, there has been a propensity for the Navy to defer maintenance, with a resulting impact both on the current workload and the amount of work that needs to be executed across ship service lives. Therefore, we used historical trends to project various scenarios of deferral to show the resulting impact. Using different assumptions on how the Navy elects to retire the backlog of deferred maintenance, we show how these

scenarios result in different required maintenance levels over the near and long term.
- **Characterize the repair and modernization industrial base capacity:** To estimate future resources for supply, we first conducted a survey of the current capacity in public- and private-sector maintenance providers, looking at both key facility inventories and labor. This task was conducted using historical trend data, current data from maintenance providers, program offices and planning activities, and interviews. To gain insight into private-sector decisions, we used interviews with industry management to better understand the incentives and disincentives for making investment decisions. These interviews included the ship repair associations and regional maintenance centers (RMCs) in Hampton Roads, Puget Sound, Jacksonville, and San Diego, as well as representatives from Huntington Ingalls Industries–Newport News Shipbuilding (HII-NNS), Continental Maritime of San Diego (CMSD), BAE, General Dynamics, Vigor Industrial, Pacific Ship Repair, and Marine Hydraulics International (MHI). These interview results are reported in Chapters Four and Five.
- **Compare the supply and the demand of resources:** Using supply and demand data and findings, we were able to identify potential misalignments. We also examined future workload demand and the adequacy of current capacity to meet this demand. For key facilities, we looked in detail at dry docks, matching availability to the expected docking requirements in CMPs and TFPs. While we could make some projections concerning expected future labor force from the Bureau of Labor Statistics and other sources to estimate capacity in public shipyards, the availability of labor and infrastructure from private-sector providers would depend primarily on choices made by private, for-profit companies.
- **Provide findings and recommendations.**

Study Limitations

Demand analyzed in this study is based partly on plans that the Navy has developed and partly on projections based on history. First, there is a routine mismatch between plan and execution. Figure 1.2 is drawn from Navy budget exhibits on public-shipyard expenditures and shows three amounts: the originally submitted President's Budget (PB) amount; the amount actually budgeted for execution, as reflected in the next year's budget submission; and the actual amount executed. In each case, the final executed amount showed growth over what was originally planned, by an average of nearly 15 percent and going as high as 20 percent. This means that, in every year depicted, the Navy ultimately added significantly to what it had originally planned to spend.

The example given here is for public shipyards, but similar trends are present in private shipyards. The history suggests that, whatever is projected, the reality is likely to be higher. There are a variety of factors that drive this observed trend; these include the underestimation of risk factors, productivity issues in the execution of activities and among workforce, and operational or budgetary factors that may drive less optimal scheduling of the maintenance.

The analysis was informed by past RAND research on materiel readiness and industrial base resources, as well as ongoing analyses that have synergies with the proposed research.[4] The results of the research were highly dependent on the availability of data and the cooperation of industrial base companies. Although all private-sector companies were available for discussions, few provided the data that were requested.

While ship repair and modernization are distinct activities, industrial base assessments have grouped ship repair and production into a single entity. This prevents us from using such materials as the Maritime Administration Report on the U.S. Shipbuilding and Repair Industrial Base to identify trends and capabilities only in the ship repair indus-

---

[4] See, e.g., Jessie Riposo, Michael E. McMahon, James G. Kallimani, and Daniel Tremblay, *Current and Future Challenges to Resourcing U.S. Navy Public Shipyards*, Santa Monica, Calif.: RAND Corporation, RR-1552-NAVY, 2017.

Introduction 7

Figure 1.2
Mismatch Between Budget and Programming

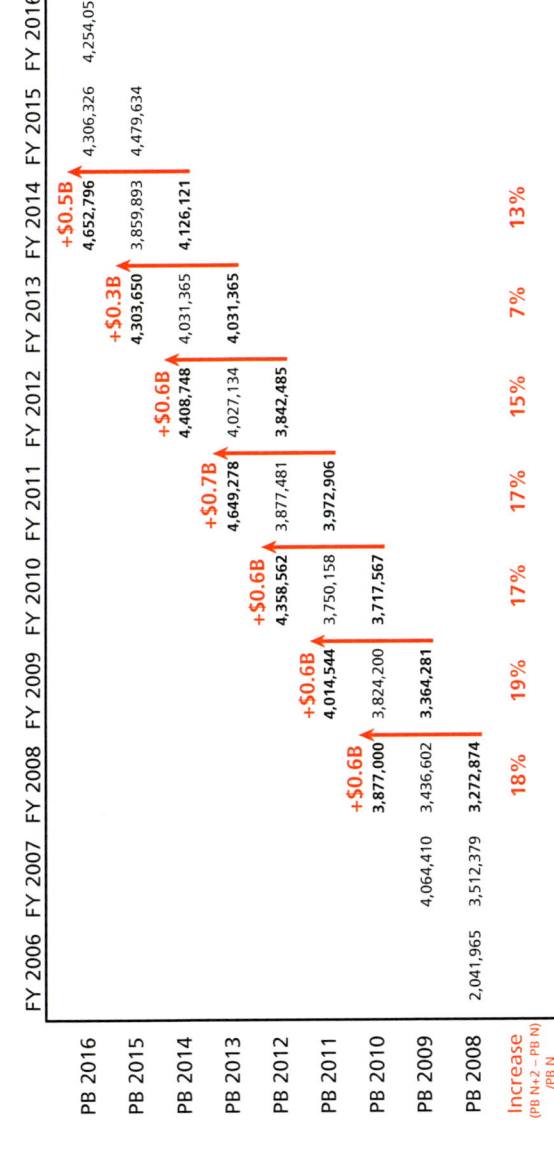

SOURCE: Author analysis of Department of the Navy, President's budget estimates, Operations and Maintenance, multiple years, FYs 2006–2016.
NOTE: The first number in each row represents the programmed amount two years before the PB was to be executed; the second number is the amount submitted as the expected execution one year before the PB was to be executed; the third number is the amount budgeted for execution in the actual year of the PB.

RAND RR1951-1.2

try.[5] The labor halls and unions that advocate for the trades do not necessarily track trades by sector, either. For example, the welders within the union may work on a variety of tasks outside U.S. Navy ship repair.

There are companies that execute both ship construction and repair services. For these companies, the resources associated with repair activities are not always easy to separate. Resources can and do move from one part of the company to the other, as needed. In addition, the production workload can affect the ability of the company to provide repair services, when production is the primary line of business. While we focus on the repair industrial base, it is not possible to isolate the providers for either nuclear or nonnuclear, surface ship or submarine, completely from the ship construction providers.

Organization of This Report

The remainder of this report is organized as follows: An overview of the demand for maintenance is provided in Chapter Two. The challenges associated with predicting future requirements and the potential alternatives are discussed. Chapter Three presents an overview of the industrial base that supports the U.S. Navy. Chapter Four discusses key findings concerning private industry's incentives and challenges in supporting ship maintenance. Mismatches between supply and demand are discussed in Chapter Five. Chapter Six discusses some potential alternatives for mitigating the challenges identified in Chapter Four.

Two appendixes support this study. Appendix A presents a look at the shipbuilding and maintenance capabilities available throughout the United States, as well as relevant details of individual shipyards. Appendix B describes the way we converted Visibility and Management of Operations and Support Costs (VAMOSC) data for use in this study.

---

[5] See Maritime Administration, "Shipyard Reports," web page, undated.

CHAPTER TWO
The Future Demand for U.S. Navy Ship Repair

RAND has done a number of studies in the past four years on ship maintenance, ranging from consideration of ship operational cycles,[1] to growth in surface ship maintenance requirements,[2] to public-shipyard maintenance,[3] to reasons for the persistent increase in Navy ship depot maintenance requirements. These were all directed toward explaining historically observed trends, trends that may be having and will continue to have an impact, but with a focus toward the past rather than the future. This study specifically looks toward the future.

The Elements of Demand

The Navy manages maintenance and modernization on all its ships throughout each ship's service life. The demand for maintenance services depends on several factors, which we will examine in detail. The first is, simply enough, the force structure. The Navy will maintain some number of aircraft carriers, submarines, surface combatants,

---

[1] Roland J. Yardley, John F. Schank, James G. Kallimani, Raj Raman, and Clifford A. Grammich, *A Methodology for Estimating the Effect of Aircraft Carrier Operational Cycles on the Maintenance Industrial Base*, Santa Monica, Calif.: RAND Corporation, TR-480-NAVY, 2007.

[2] Robert W. Button, Bradley Martin, Jerry M. Sollinger, and Abraham Tidwell, *Assessment of Surface Ship Maintenance Requirements*, Santa Monica, Calif.: RAND Corporation, RR-1155-NAVY, 2015.

[3] Riposo et al., 2017.

amphibious ships, and auxiliaries intended to satisfy presence and warfighting requirements. All are built with an expected service life; all will require maintenance and modernization throughout their service lives. The second component is what actually must be accomplished on these ships and submarines to reach service life. These are captured in documents developed by NAVSEA, including TFPs and CMPs. These two types of documents describe the work required at different stages of a ship's life, including periods requiring dry-docking and major modernizations. These plans are tailored to the individual platforms to support unique platforms and installed systems maintenance and certification requirements.

The 30-Year Shipbuilding Plan Defines Future Force Structure

To help communicate future shipbuilding needs with the industrial base and Congress, the Navy produces an annual report, *Report to Congress on the Annual Long-Range Plan of Construction of Naval Vessels*.[4] This report describes the force structure necessary to "fulfill the Navy's essential combat missions at an acceptable level of risk."[5] Also presented is a 30-year construction plan that identifies the numbers and types of ships that the Navy intends to buy each year to achieve force structure goals. In addition to platform build plans, the report depicts the planned ship and submarine decommissionings. Table 2.1 shows the expected annual build rate for different ship classes.

The build plan and platform decommissionings result in a force structure depicted in Figure 2.1. Based on the most-recent Navy plans, the future force structure reflects an increasingly large number of small surface combatants over time. In general, however, the current plan does not show significant changes in force structure, with new units replacing old ones and numbers remaining in a narrow range over time. This does not take into account the Navy's current assessment that it, in fact, requires a larger force structure to meet its projected missions.

---

[4] The report is prepared each FY by the Office of the Chief of Naval Operations (OPNAV), Deputy Chief of Naval Operations (Integration of Capabilities and Resources). See Office of the Chief of Naval Operations, 2016, for an example.

[5] Office of the Chief of Naval Operations, 2016, p. 3.

Table 2.1
Annual New Construction Plan

Ship Class	2016	2017	2018	2019	2020	2021	2022	2023	2024	2025	2026	2027	2028	2029	2030
Aircraft carrier			1					1					1		
Large surface combatant	2	2	2	2	2	2	2	2	2	2	2	2	2	2	2
Small surface combatant	3	3	3	2	3	3	3	3	3	3					1
Attack submarine	2	2	2	2	2	1	2	2	1	2	1	1	1	1	1
Ballistic missile submarine						1			1		1	1	1	1	1
Amphibious warfare ship	1	1			1		1	1	2	1	1	1	2	1	1
Combat logistics force	1		1	1	1	1	1	1	1	1	1	1	1	1	1
Support vessel		2	1	2	1	1	2	3	2	1			1	1	2
Total new construction plan	9	10	10	9	10	9	11	13	12	10	6	6	9	7	9

Table 2.1—Continued

Ship Class	2031	2032	2033	2034	2035	2036	2037	2038	2039	2040	2041	2042	2043	2044	2045
Aircraft carrier			1					1					1		
Large surface combatant	2	2	2	2	2	2	2	3	3	3	2	3	2	3	2
Small surface combatant		1	1	1	2	2	2	3	4	4	4	4	4	2	3
Attack submarine	1	1	1	1	1	2	2	2	2	1	2	1	2	1	2
Ballistic missile submarine	1	1	1	1	1										
Amphibious warfare ship	1	1				1				2		1		2	1
Combat logistics force	1	1	1										1		2
Support vessel	2	2	2	1											
Total new construction plan	8	9	9	6	6	7	6	9	9	10	8	9	10	8	10

SOURCE: OPNAV, 2016.
NOTE: Blank cells = nothing delivered that year.

Figure 2.1
Future Naval Ship Force Structure

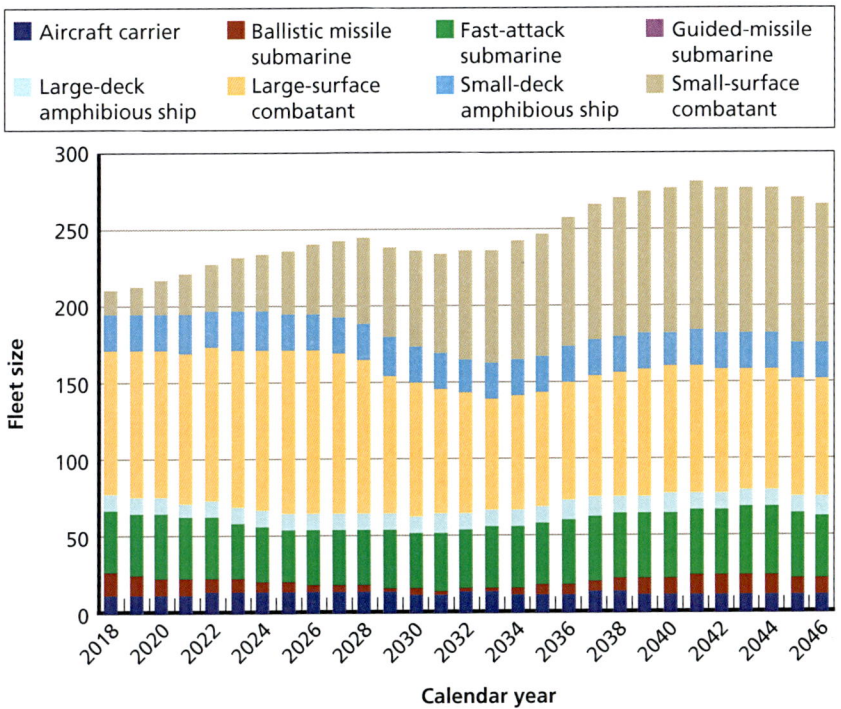

SOURCES: OPNAV, 2016; NAVSEA, *Surface Maintenance Engineering Planning Program Class Depot Maintenance Technical Foundation Paper*, Washington, D.C., various years and for different classes.
RAND RR1951-2.1

Such an expansion, in the near term, would require both additional new ships and the retention of older ones, involving work to extend the service lives of some platforms. Both would add to maintenance demand beyond the current plan. It should be noted that substantial work is generated in decommissioning of ships, especially for nuclear-powered vessels.

To ensure the highest operating capabilities, ships require both a preplanned defined amount of maintenance during their service lives and necessary but variable condition-based maintenance. The life-cycle plan then is based, with varying degrees of formality, on an assessment of technical requirements, which takes into account the ship's design, its operating environment, key systems, and its mission. For some ship classes, a complete TFP has been generated to capture maintenance needed at every juncture in a ship's life. These TFPs are developed by the life-cycle platform managers with the technical warrant holders and planning activities within NAVSEA and represent a systematic effort to codify the actual work needed to ensure that ships are operable in the short term and reach service life in the long term. TFPs include periods of dry-docking, as well as a factor for aging and a factor for modernization.[6] The resultant maintenance work in the TFP is scheduled into a plan with dedicated ship availabilities at phased intervals in the service life, some involving dry-docking the ship for hull, tank, and other work with longer durations and other availabilities without a dry-docking. As an example, Figure 2.2 shows the maintenance required by the *Arleigh Burke*–class (DDG-51) Flight I and II TFPs. The horizontal axis shows the intervals of major availabilities in the phased service life, and the vertical axis shows the notional man-days in the plan for each interval.

As can be seen in Figure 2.2, the total work for an individual DDG-51 is phased to different points in the ship's life, with a large proportion of the work occurring in the midlife maintenance period. Within each availability, a TFP may specify the type of work required and the time required within each. This is noted by the SWLIN. Figure 2.3 shows an example from the DDG-51 TFP of man-days divided by the SWLIN in particular availabilities. The figure shows that in SRA 3-1 and 3-2, SWLIN codes 1X (hull structure) through 8X (services), a certain number of days would be planned through the long-range maintenance system (LRMS) and then adjusted for aging and modernization alterations to arrive at a number of man-days for each SWLIN.

Although TFPs are authoritative in that they describe the maintenance basis, they are not the sole determinants of what work will

---

[6] NAVSEA, *Technical Foundation Paper for DDG 51 Class*, Washington, D.C., 2012b; NAVSEA, *Surface Maintenance Engineering Planning Program LPD 17 Class Technical Foundation Paper*, Washington, D.C., May 23, 2012a.

Figure 2.2
DDG-51 Work Distribution in the TFP

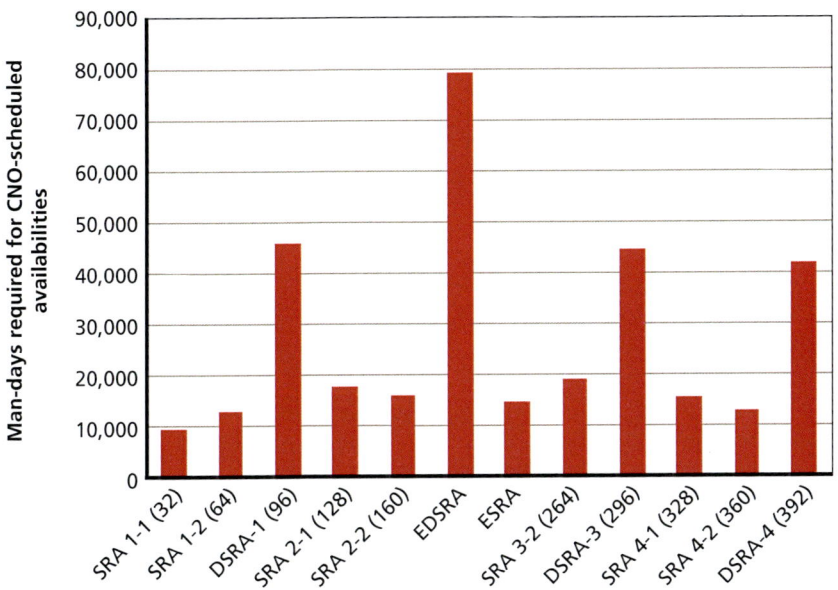

SOURCE: NAVSEA, 2012b.
NOTES: The parentheses indicate months into the ship's service life. CNO = Chief of Naval Operations; SRA = Selected Restricted Availability; DSRA = Docking Selected Restricted Availability; ESRA = Extended Selected Restricted Availability; EDSRA = Extended Docking Selected Restricted Availability.
RAND RR1951-2.2

be completed in an availability. The actual condition of the ship may drive greater or lesser degrees of maintenance to be performed in a given availability, as may available funding. However, it is reasonable to treat TFPs as reliable planning guides for scheduling and the complete maintenance requirement that looks toward future maintenance workloads for planning purposes, with the caveat that the projected work may be somewhat higher. For purposes of analysis, we assumed that work required, but not done when scheduled by the TFP, will eventually require completion and remains in the overall maintenance demand. We will examine the possible impacts of deferral on mainte-

Figure 2.3
SWLIN Distribution in Representative DDG-51 Availabilities

SRA 3-1	100	200	300	400	500	600	700	800	Totals
LRMS notional	646	1,073	371	2,698	2,162	2,032	301	2,908	12,191
TFP total	791	1,528	405	2,945	2,382	2,227	327	3,141	13,745

SRA 3-2	100	200	300	400	500	600	700	800	Totals
LRMS notional	1,807	1,073	364	2,888	2,249	2,032	1,424	4,323	16,160
TFP total	2,144	1,528	397	3,151	2,465	2,225	1,541	4,669	18,118

SOURCE: NAVSEA, 2012b.
NOTES: These numbers represent only two availabilities out of several in the DDG-51's service life. Numbers are rounded for display purposes and might not add to displayed totals. The TFP total includes LRMS, aging, and alterations. The SWLIN series refer to the following: 100 series = hull structure; 200 series = propulsion; 300 series = electrical plant; 400 series = command surveillance; 500 series = auxiliary systems; 600 series = outfitting and furnishing; 700 series = armaments; 800 series = assembly and support.
RAND RR1951-2.3

nance demand later in this chapter, but we begin with the assumption that maintenance—deferred and otherwise—remains a requirement.

While not every ship has a TFP, all operate within a CMP that allows for forecasting the required maintenance and required resources for each ship across a number of years. For ship classes with a TFP, the TFP is used as the CMP. For those without a TFP, less detailed CMPs are developed. These are formulated at the delivery of ship classes and are updated periodically as more is understood about ship maintenance requirements. Figure 2.4 is an example of CMP workload, in this case for the *Virginia*-class (SSN-774) submarine. This particular workload is noteworthy because the plan itself changed significantly—nearly doubling in overall requirement from the time of the lead ship's delivery. Although the plan has continued to evolve, with only minor changes since the large increases in 2008 and 2009, this jump would create great uncertainty in workload planning for the shipyards that provide life-cycle maintenance and the sponsors that provide maintenance resources. The maintenance costs were poorly understood or possibly understated during the acquisition process, with a large impact on future resource management.

Figure 2.4 shows that the original plan (2004–2008) called for eight smaller maintenance periods, for a total of roughly 450,000 man-days across the life of each SSN-774. Following the rebaselining in 2009, four larger maintenance periods were planned, totaling more 800,000 man-days of maintenance over the life of a single SSN-774.

Deferral Trends

Every trend discussed so far is notional and based on class plans. In reality, the Navy frequently elects to defer maintenance—that is, not perform it at the scheduled time, usually with the intention of performing it later. This is generally not done for submarine maintenance related to diving safety or to components in nuclear power plants and

Figure 2.4
Virginia-Class (SSN-774) Maintenance Plan

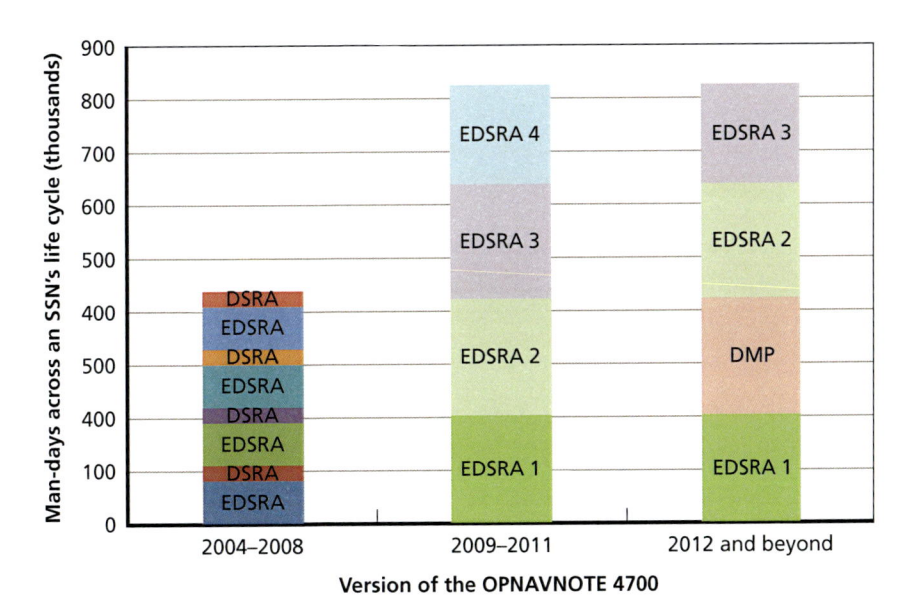

SOURCE: Office of the Chief of Naval Operations, *Representative Intervals, Durations, and Repair Man-Days for Depot Level Maintenance Availabilities of US Navy Ships*, OPNAVNOTE 4700, Washington, D.C., 2004–2016.
NOTE: DMP = depot modernization period.
RAND RR1951-2.4

18   A Strategic Assessment of the Future of U.S. Navy Ship Maintenance

auxiliaries. It is routinely done for a variety of reasons in surface combatants and amphibious ships. Indeed, historical programming guidance has routinely resourced only 80 percent of the annual maintenance requirement for surface ships.[7]

When necessary work, modernization, or availability is deferred, other availability schedules are affected. This may lead to larger outyear requirements. Figures 2.5 and 2.6 show, historically, the gap between maintenance required in TFPs for DDG-51 and the CMP for *Ticonderoga*-class (CG-47) guided-missile cruisers. In the case of DDG-51, from 2003 to 2010, the prescribed maintenance falls behind the maintenance actually performed, as reflected in the Navy

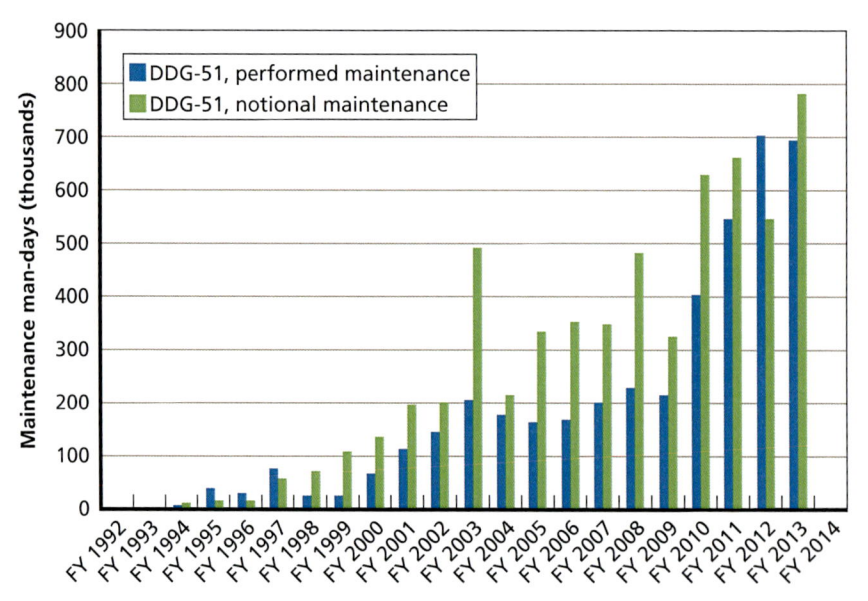

Figure 2.5
DDG-51 Performed Versus TFP-Directed Maintenance

SOURCES: NAVSEA, 2012b; and annual expenditure calculations from the Navy Center for Cost Analysis's VAMOSC database, FYs 1992–2014.
RAND RR1951-2.5

---

[7] Phillip Balisle, *Final Report: Fleet Review Panel of Surface Ship Readiness*, U.S. Fleet Forces Command and U.S. Pacific Fleet, February 26, 2010.

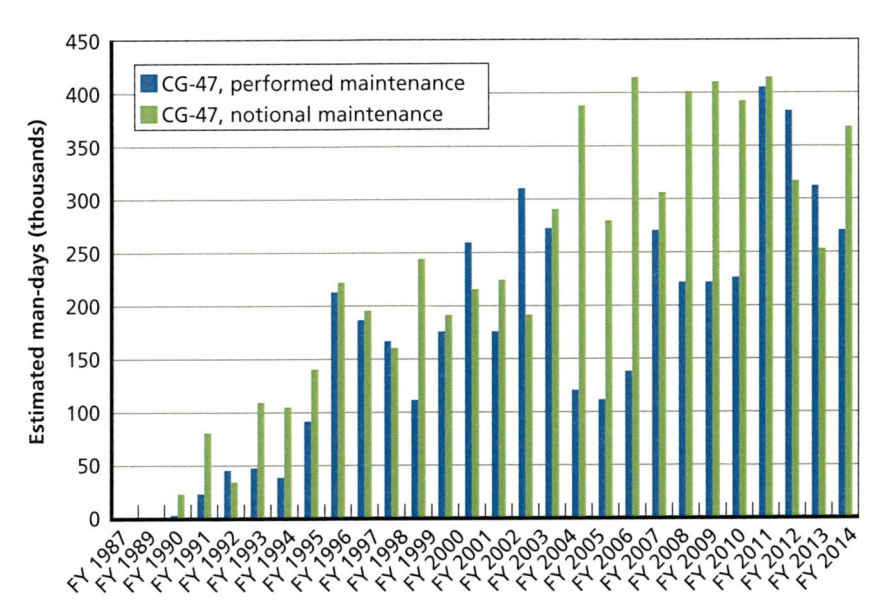

Figure 2.6
CG-47 Performed Versus CMP-Directed Maintenance

[Bar chart showing estimated man-days (thousands) by fiscal year from FY 1987 to FY 2014, comparing CG-47 performed maintenance (blue) and CG-47 notional maintenance (green).]

SOURCES: Office of the Chief of Naval Operations, 2014; annual expenditure calculations from the Navy Center for Cost Analysis's VAMOSC database, FYs 1987–2014.
NOTE: The CG-47 expenditure in man-days is labeled estimated because VAMOSC data are in dollars and a conversion factor was used to calculate maintenance man-days.
RAND RR1951-2.6

VAMOSC data, with FY 2018 dollars converted to man-days.[8] Since 2011, this relationship has reversed, and work actually performed has been essentially equivalent to, or somewhat higher than, what the TFP would direct.

In the case of the CG-47, from 2003 to 2009, the gap between what the CMP directed and what was actually accomplished is dramatic: In some cases, less than a quarter of what was scheduled to be accomplished was actually performed. This was reversed in 2011, for at least two years, with the performed maintenance exceeding the

---

[8] See Appendix B for a description of how VAMOSC data were converted for use in this study.

annually directed maintenance. There are several possible explanations for this. One is that the future of the class was not determined until recently, with a presumption of and, to a degree, preference for decommissioning the ships as they approached expected service life. One possible explanation for the change in maintenance executed was the Navy's reaction to a 2010 report, a Fleet Review Panel effort that criticized the surface-ship Navy for poor material condition and maintenance deferral.[9] Cruisers, in particular, were shown to have poor levels of material readiness, evidenced by poor inspection performance and difficulty meeting operational commitments. In any case, Congress has consistently resisted decommissioning these ships and, in fact, has directed that the ships be kept in commission and modernized. The addition of this substantial volume of maintenance to reset the CG-47 material condition and modernize the ships reflects a decision to retain the hulls for a longer service life; this substantial maintenance will overlay the planned fleet maintenance. The execution of this work will likely be a national effort that will involve the shipbuilding base, given the scope of the effort.

Maintenance Funding Across the Future Years Defense Plan

Given previous occasions where less maintenance was performed than required, an examination of a five-year period in the current Future Years Defense Plan (FYDP) yields the somewhat surprising finding that programmed and budgeted maintenance actually exceeds what the CMP or the TFP would have dictated (Figure 2.7).[10] The profile is apparent in both public and private shipyards (Figure 2.8) and is in some cases significant. What remains to be seen is whether the industrial capacity exists to accomplish this funding profile. Certainly, there needs to be flexibility for the Navy to assign this work outside homeports, when possible, and use the full range of naval shipyards, private repair providers, and shipbuilder resources.[11]

---

[9] Balisle, 2010.

[10] See Department of the Navy, budget materials for fiscal year 2018, web page, undated.

[11] Performing work outside homeports would require changes in personnel management policy and practice.

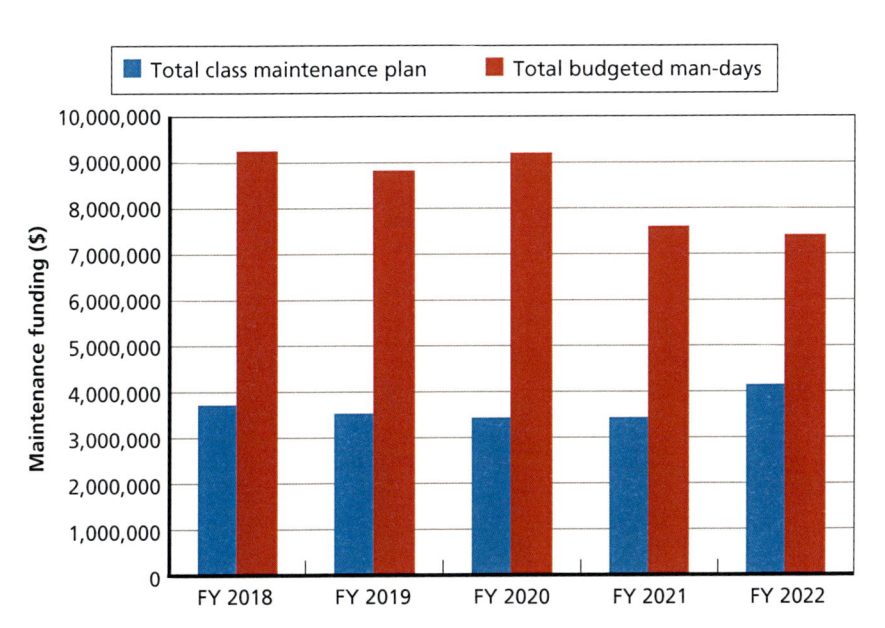

Figure 2.7
Difference Between CMP and Actual Budgeted Man-Days

[Bar chart showing Total class maintenance plan (blue) and Total budgeted man-days (red) for FY 2018 through FY 2022. Y-axis: Maintenance funding ($), from 0 to 10,000,000.]

SOURCES: Office of the Chief of Naval Operations, 2014; Department of the Navy, 2017.

RAND RR1951-2.7

This mismatch may reflect the previously noted historical gap between programmed and executed funding. However, the gap goes beyond what is in the budget versus what is in the program objective memorandum to depicting a difference between the technical requirement and the level of maintenance effort that fleets and NAVSEA expect to carry out. This could possibly reflect a commitment by the Navy to recover maintenance as a result of earlier deferrals; the Navy has officially committed to investment in material wholeness.[12] That action would be a reversal of previous and persistent trends.

---

[12] Bill Moran, Adm., Vice Chief of Naval Operations, testimony before the Senate Armed Services Committee, February 8, 2017.

Figure 2.8
Difference Between CMPs and Programmed Levels in Private- and Public-Shipyard Availabilities

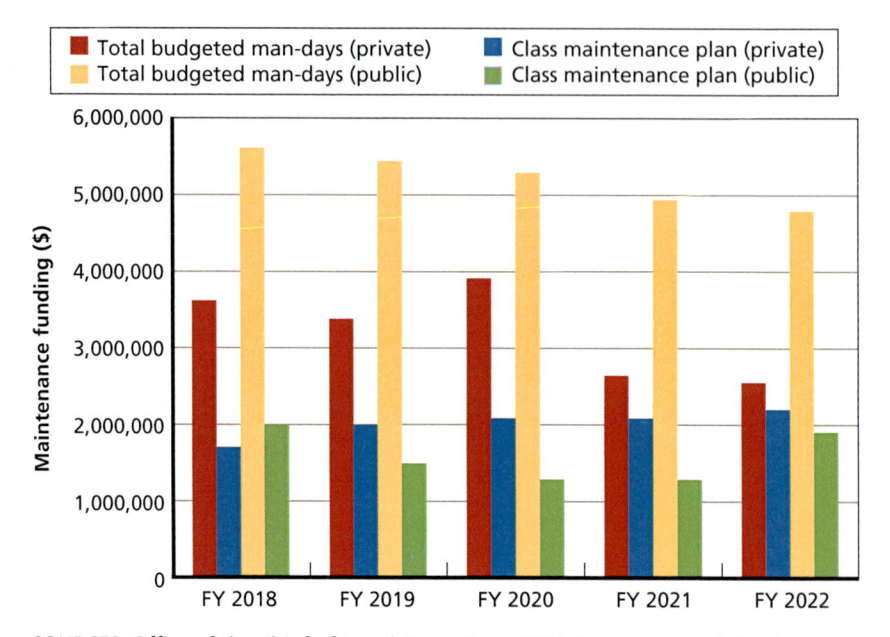

SOURCES: Office of the Chief of Naval Operations, 2014; Department of the Navy, 2017.
RAND RR1951-2.8

Resulting Demand Trends If the Plans Are Executed Without Deviation

If the Navy executes its 30-year shipbuilding plan, including the stated ship decommissioning plan and completed maintenance as specified in CMPs, there will be periods of high and low demand, with a variation of approximately a million man-days of maintenance every year.[13] Figure 2.9 shows the expected annual changes for future workload, based on fleet inventory by type of platform and the prescribed man-days of maintenance in the TFP or other planning information. This omits carrier refueling complex overhauls (RCOHs), which are conducted only at HII-NNS using, largely, shipbuilder labor as opposed to

---

[13] Office of the Chief of Naval Operations, 2016.

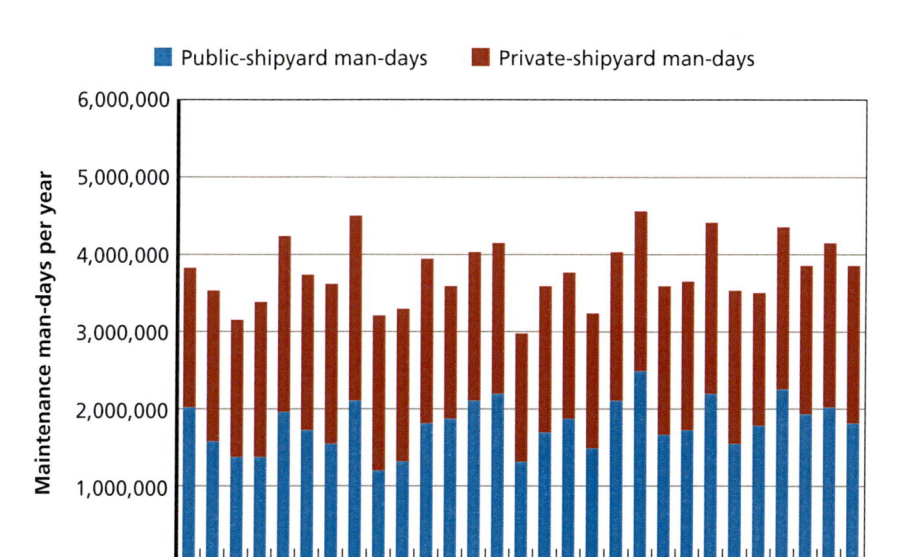

Figure 2.9
Planned Maintenance Workloads

■ Public-shipyard man-days   ■ Private-shipyard man-days

[Bar chart showing maintenance man-days per year from 2018 to 2046, with public-shipyard and private-shipyard man-days stacked. Y-axis ranges from 0 to 6,000,000 maintenance man-days per year.]

SOURCES: Office of the Chief of Naval Operations, 2014; Office of the Chief of Naval Operations, 2016.
RAND RR1951-2.9

a maintenance at a NSY or and private-sector repair facility. According to 1B4B depot maintenance account funding data, all maintenance conducted on surface ships and half of the nonnuclear maintenance conducted on aircraft carriers are conducted in private shipyards.[14] Figure 2.9 shows private-shipyard maintenance as relatively stable across time overall (but with large variations at the port level, to be discussed later), with the larger perturbations of overall being due to variations in public-sector workload. The impact of these variations is significant in both sectors because the total variation in the public-

---

[14] The Navy resources maintenance and repair via a ship depot maintenance account—budget code 1B4B (Budget Activity: 1 Operating Forces, Activity Group: B Ship Operations, Detail by sub-activity: 4B Ship Maintenance); see Department of the Navy, FYs 2000–2018.

sector NSYs, for example, is spread across only four shipyards, while the variation is spread across more providers in the private sector but primarily only in four homeports.[15]

Labor Force Demands in the Private Sector

TFPs specify not only a total number of expected man-days but also the man-days expected within particular maintenance areas, as specified in SWLINs. Figure 2.10 shows a future distribution of SWLINs in private-shipyard availabilities projected by fleet composition in the Long-Range Shipbuilding Plans and using the TFPs. The distribution does not vary significantly across time. The types of work needed, by SWLIN, in 2016 are very similar to the types of work required decades later, as predicted by the TFPs and maintenance plans. This is not to say that the assignment or execution of the work will not change; it does suggest that persistent factors, such as corrosion, complex interfaces, and distributed systems, will remain the same and that the labor trades required today will likely be required far into the future.

Results When Scheduled Maintenance Is Deferred

We now shift from presenting plans or executed history to applying historical trends to future projection. The maintenance demands that we expect if the 30-year shipbuilding plan is executed as planned are predictable and, while cyclical, appear to be manageable. However, as we have seen, for at least some ship classes, there is a widespread practice of deferring maintenance. It is important to assess the impact that this practice might have if deferral trends are continued. Not all deferred maintained will necessarily be recovered. Decks that are not preserved are not preserved twice in subsequent availabilities. The corrosion and damage may have become more extensive, but work not accomplished in one availability is more likely to be spread across the life of the ship

---

[15] This does not include the decommissioning and disposal expenses associated with CVN-68 decommissioning. Seven of these will occur in the 30-year shipbuilding plan time frame. These occur at five-year intervals, although there may be variations in execution. In the case of CVN-65, the work was done by the builder's yard, and it appears that the major impact on the maintenance industrial base will be an additional impediment on using builder's yards for ship maintenance.

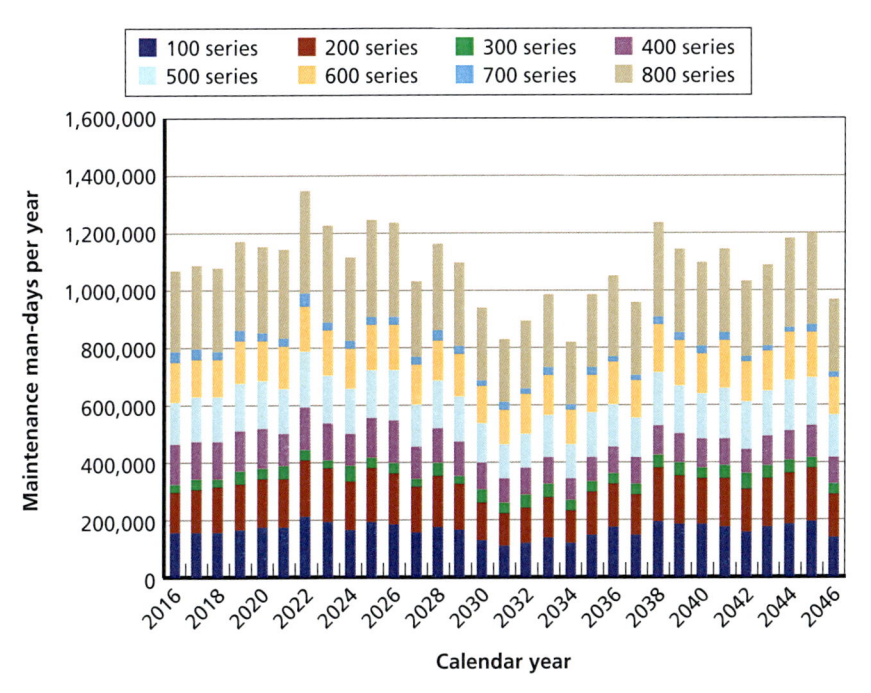

Figure 2.10
Private-Sector Equipment Distribution

[Stacked bar chart showing maintenance man-days per year from 2016 to 2046, with categories: 100 series, 200 series, 300 series, 400 series, 500 series, 600 series, 700 series, 800 series. Y-axis ranges from 0 to 1,600,000.]

SOURCES: Office of the Chief of Naval Operations, 2016; NAVSEA, 2012a; NAVSEA, 2012b.
NOTE: 100 series = hull structure; 200 series = propulsion; 300 series = electrical plant; 400 series = command surveillance; 500 series = auxiliary systems; 600 series = outfitting and furnishing; 700 series = armaments; 800 series = assembly and support.
RAND RR1951-2.10

rather than simply added to the next. However, it generally is true that the same work accomplished later will be expensive, if only because of unaddressed problems leading to unknown growth. The Navy itself applies a 6 percent annual "fester factor" to deferred work.[16]

To explore the possible impacts of deferrals, we looked at the historical deferral figures for the DDG-51. We considered the DDG class for several reasons. First, it is the most numerous ship class in the Navy,

---

[16] "Burke: $2 Billion Backlog in Surface Ship Maintenance Hard to Dig Out Of," InsideDefense.com, March 22, 2013.

with a wide temporal spread between the oldest and the newest ships. Second, it has an approved TFP, which specifies the amount of maintenance that should have occurred on an annual basis.[17] As we have seen, there is a mismatch that indicates that less was actually done than called for in the TFP. The differences between what the TFP called for and the maintenance actually performed we defined as *deferred*. We then used these historically observed rates to make projections about how deferral might affect the entire Navy if the deferral rates are similar.

Figures 2.11 through 2.13 contain the same basic items. These are

- scheduled days: amount of maintenance that should be performed based on CMPs
- historical deferred days: amount of deferred maintenance from before 2018 that needs to be performed
- projected deferred days: amount of maintenance that is expected to be deferred from the scheduled days, based on historical trends
- delayed deferred days: amount of extra maintenance that would be required if deferred maintenance is not performed on a regular schedule.

In each of these cases, we applied an annual 6 percent growth rate on the unaccomplished maintenance.

Figure 2.11 shows the annual additions to required ship maintenance under the following deferral conditions: the current known deferral is carried forward; the Navy defers some additional maintenance each year, as it has historically done; and each year the Navy tries to retire some deferments from the previous year. To paraphrase, the Navy does not necessarily accomplish the maintenance called for in the TFPs and the CMPs but persistently attempts to retire the maintenance backlog. When done, there is an annual 15–20 percent difference between what the TFP or the CMP would have directed and the amount of maintenance actually required with deferral taken into account.

---

[17] NAVSEA, 2012b.

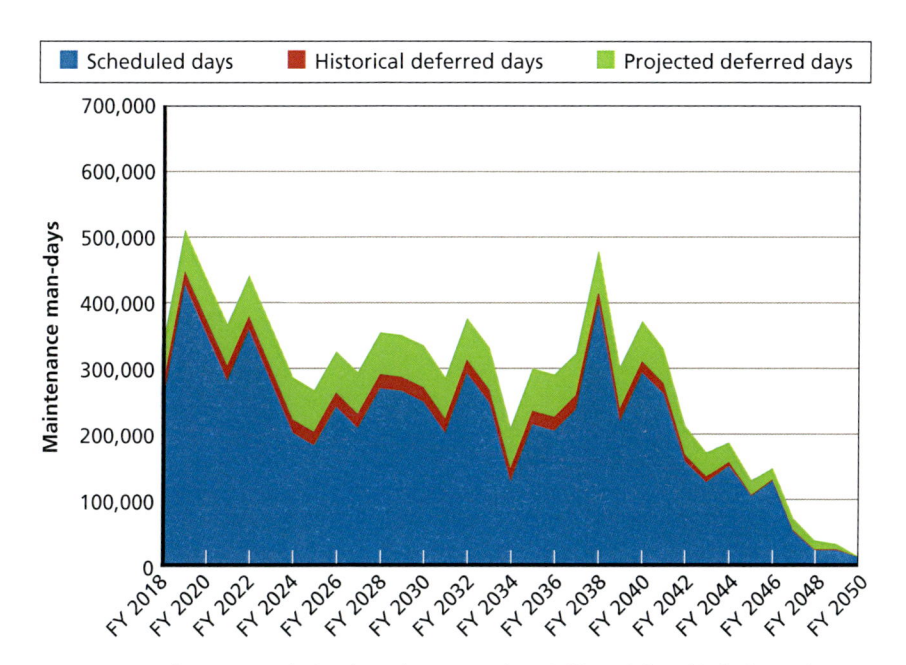

Figure 2.11
Added Maintenance with Regular Recovery of Deferred Maintenance

SOURCES: Author extrapolation based on CMPs (see Office of the Chief of Naval Operations, 2014) and Navy RMC's Navy Maintenance Database.
NOTES: There are no delayed deferred days in this scenario. The deferral conditions are the following: current known deferral is carried forward; the Navy defers some additional maintenance each year, as it has historically done; and each year the Navy tries to retire some deferments from the previous year.
RAND RR1951-2.11

However, the pattern of deferral might suggest that, rather than a regular effort to recover lost days, the Navy recovers maintenance over time. Figure 2.12 depicts the results when the Navy allows deferred maintenance to grow for an additional five years before attempting to retire it. Since this would amount, essentially, to funding maintenance to no more than previously required levels through the FYDP, this is not implausible under various fiscal circumstances. Here, the difference between scheduled and required maintenance is 20–40 percent annually, suggesting a large long-term cost for this behavior.

The case of the CG-47 suggests that even longer-term deferrals may occur, which we need to examine. Figure 2.13 shows the impact

Figure 2.12
Added Maintenance with Five-Year Deferral of Maintenance Backlogs

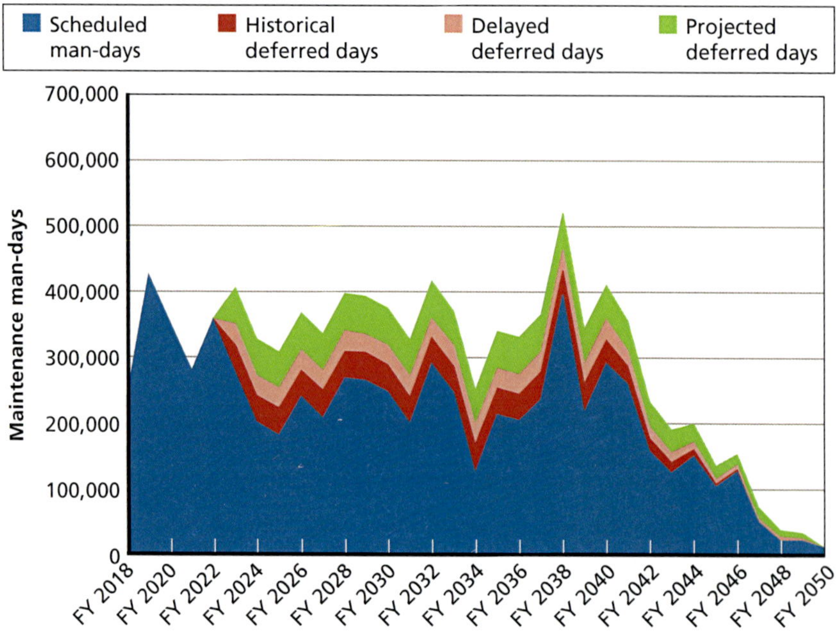

SOURCES: Author extrapolation based on CMPs (see Office of the Chief of Naval Operations, 2014) and the Navy RMC's Navy Maintenance Database.
RAND RR1951-2.12

if the deferral is as long as ten years before the backlog is addressed. In this case, the out-year impact is dramatic, with the deferred maintenance at times reaching half the scheduled amount. It is worth noting that the impact actually resembles the mismatch between the previously programmed and executed budged amounts discussed earlier in the chapter.

Taken together, these three cases demonstrate that deferral of maintenance because of funding or schedule, or any reason other than clear identification that material condition justifies the foregoing of some maintenance, results in significant impacts on the required levels in the out-years.

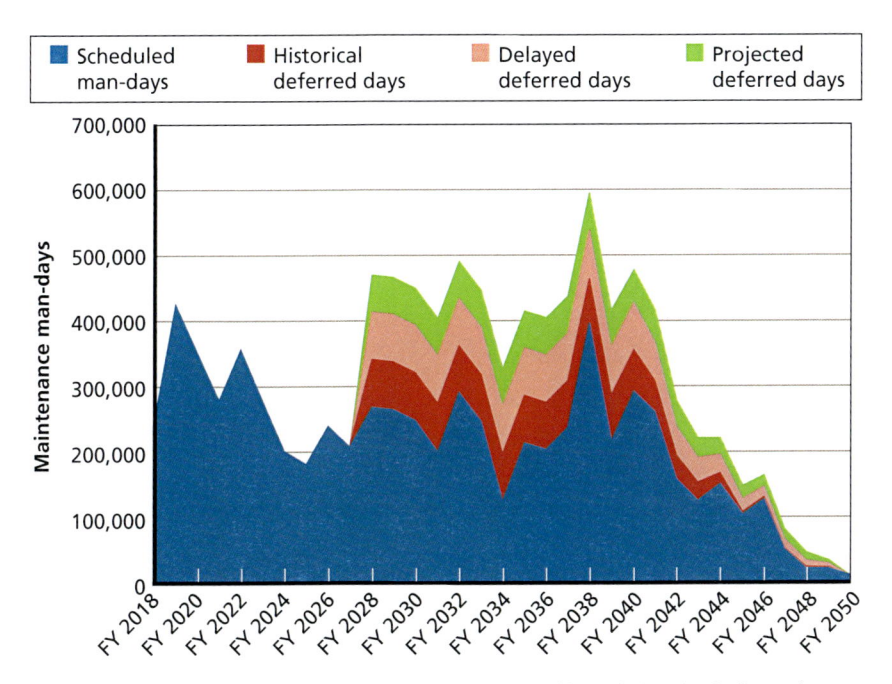

Figure 2.13
Added Maintenance with Ten-Year Deferral of Maintenance Backlogs.

SOURCES: Author extrapolation based on CMPs (see Office of the Chief of Naval Operations, 2014) and the Navy RMC's Navy Maintenance Database.
RAND RR1951-2.13

Maintenance Demand Conclusions

If the 30-year shipbuilding plan is executed as planned and if the Navy makes a consistent effort to comply with its CMPs, long-range, future maintenance workload, based on the current Long-Range Shipbuilding Plan for fleet inventory, will remain at least at current levels, with historical trends suggesting that higher maintenance levels are likely. This projection applies in both the public and the private sectors.

The type of workload (and, hence, the labor skills expected to be required) is also not likely to change, with the distribution by SWLIN appearing consistent in the decades to come. Given that the operating requirements of ships and submarines will not change significantly and the nature of naval ship design and construction, this is not surprising. This indicates that trade-labor demand by skill will continue to require similar skills to current trades and also new skills to maintain fiber optics systems, photonics, control system software, and power electronics. Demands for facilities—in particular, dry docks—will be significant and, at times, overstress available dry docks by port, but the dry-dock demand may be accomplished by allowing coast-wide bidding for dry-dock availabilities. Our analysis indicates that the dry dock demand predicted currently for the LCS-1 and LCS-2, when analyzed by homeport, is not executable within available facilities.

Deferral of maintenance actions will complicate the management of maintenance demands. Deferrals occur for a variety of reasons—including funding shortfalls, scheduling demands, and capacity shortfalls—and it is unrealistic to simply insist that deferrals not occur. However, it is important to understand the impact. Our historical data show that the Navy has a tendency to defer maintenance on the two classes of surface ships examined (DDG-51 and CG-47). Our analysis of the FYDP shows, conversely, that the Navy is planning to spend more than what the technical requirements would have dictated. Our models indicate that this is likely due to an attempt to recover lost maintenance and that the impact on out-year requirements gets more severe the longer the maintenance is deferred. At a minimum, if maintenance is to be deferred, there should be a conscious effort to retire the deferrals on a consistent basis.

The Navy largely manages demand separately along public- and private-sector providers by platform, even though the supporting funded account, the 1B4B depot maintenance account, is a single account. This approach may need to shift to better use private-sector capacity and expertise to support the NSYs regionally in some areas, such as submarine tank preservation and nonnuclear carrier work in large modernization alterations—for example, to permit the NSYs to focus on core workload that might not be supported externally. This may be an essential focus area if planned fleet-force structure commences generating more demand from the shipbuilders that currently augment the ship repair providers, as well as in the midterm, when additional platforms might be delivered.

In the next chapter, we describe the available capability and capacity for meeting these maintenance demands. There is evidence that public shipyards do not have sufficient capacity to effectively meet the demand for nuclear submarine availabilities. Indeed, the Navy has elected to not induct at least one submarine into availability, leaving it unable to operate as a submarine, because of insufficient capacity in public shipyards to accomplish the maintenance.[18] While there is certainly some logic to deferring induction when there simply is not capacity for accomplishing the maintenance, this amounts to a multi-year deferral—the consequence of which may be the growth in requirements presented earlier in the chapter.

---

[18] Megan Eckstein, "Ingalls Shipbuilding Launches First Ship Since Destroyer Program Restart," *U.S. Naval Institute News*, March 30, 2015.

CHAPTER THREE

The U.S. Navy Ship Maintenance Industrial Base

The U.S. Navy ship maintenance industrial base consists of a number of both public and private providers. The public sector is primarily focused on providing maintenance services to nuclear-powered ships, while the private sector is primarily focused on providing maintenance services to nonnuclear ships—the exceptions are HII-NNS, which performs RCOH of nuclear aircraft carriers, as well as maintenance services on nuclear submarines, and General Dynamics Electric Boat, which provides maintenance services to nuclear submarines.

The Department of the Navy has a large degree of control over the capabilities that are currently and will be provided by the public sector. The Navy determines the mission and function of the NSYs and support organizations. The Navy also establishes the composition and level of workforce required to accomplish the organizations' missions and identifies and makes the investments required to ensure that the necessary capabilities are provided. The Navy has less control over the capabilities that will be provided by the private sector. The private sector responds to market forces, which, in some cases, the Navy can influence.

This chapter discusses the capabilities currently offered by the ship repair industrial base and identifies trends in the industrial base that may shape how the Navy can affect the industrial base of the future. Readers interested in the capabilities available in regions throughout the United States or in individual shipyards should refer to our fuller descriptions in Appendix A.

Defining the Maintenance Industrial Base

The Navy ship maintenance industrial base consists of the companies, organizations, people, materials, processes, and facilities required to ensure that Navy ships reach their expected service lives and can perform their required missions. Here, we focus on identifying and describing the shipyards that currently support Navy warships. While other providers and suppliers are of equal importance to the United States' ability to maintain naval ships, the capacity of the industrial base is most limited by the number of heavily facilitated shipyards. Industry representatives emphasized the significant barriers to entry for new companies, such as obtaining necessary licensing and environmental clearances.

Public Shipyards

The Navy owns and operates four public shipyards. The four NSYs conduct maintenance primarily on nuclear-powered ships, including aircraft carriers, fast-attack submarines, guided missile submarines, and ballistic missile submarines. The four public shipyards are located in different locations around the country:

- PNSY, in Kittery, Maine
- NNSY, in Portsmouth, Virginia
- PSNS & IMF. in Bremerton, Washington
- PHNS, outside Honolulu, Hawaii.

While the last two NSYs are configured with a regional role and with an integrated intermediate maintenance facility command structure, all four NSYs conduct a substantial volume of maintenance away from the main shipyard to support the execution of work at vessels' homeports, for emergent work; to augment other NSY workforces, and to conduct maintenance on forward-positioned naval forces overseas. Figure 3.1 shows the locations of the four shipyards, as well as the many locations that each shipyard supports around the world.

These shipyards perform the most-complex maintenance that the Navy requires, including most depot-level and some intermediate-level

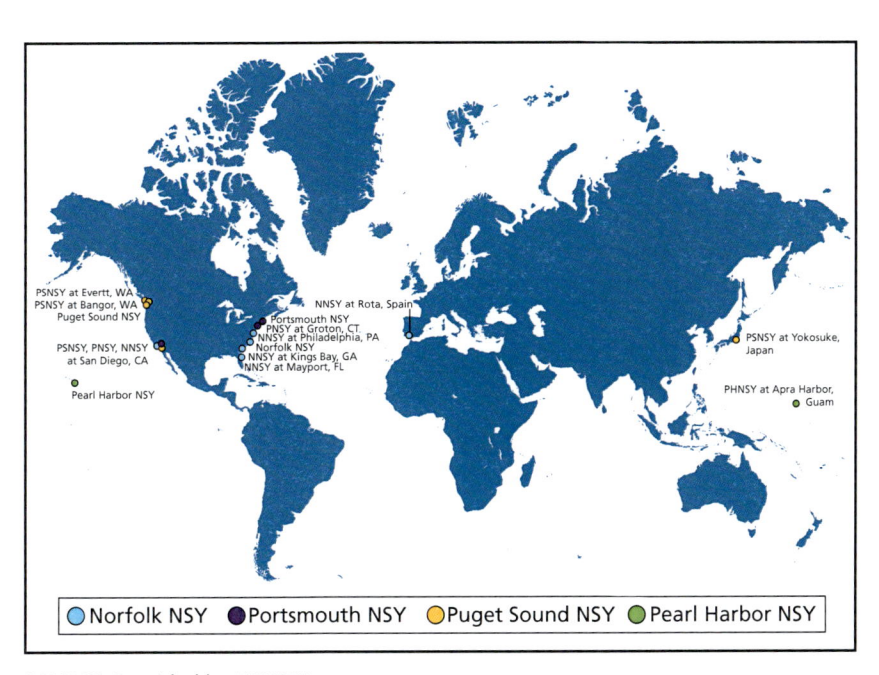

Figure 3.1
Public Shipyard and Support Locations

PSNSY at Everett, WA
PSNSY at Bangor, WA
Puget Sound NSY
PSNSY, PNSY, NNSY at San Diego, CA
Pearl Harbor NSY
NNSY at Rota, Spain
Portsmouth NSY
PNSY at Groton, CT
NNSY at Philadelphia, PA
Norfolk NSY
NNSY at Kings Bay, GA
NNSY at Mayport, FL
PSNSY at Yokosuke, Japan
PHNSY at Apra Harbor, Guam

○ Norfolk NSY ● Portsmouth NSY ○ Puget Sound NSY ○ Pearl Harbor NSY

SOURCE: Provided by NAVSEA.
RAND RR1951-3.1

life-cycle maintenance and modernization of SSBNs, SSGNs, SSNs, and CVNs. The shipyards also perform refueling of SSNs and SSBNs; life-cycle sustainment, refueling, and conversion of moored training ships (MTSs), which currently are all former SSBNs, although the next MTSs to be converted will be retired SSNs; and inactivation, reactor compartment disposal, recycling (IRR) of SSNs, SSBNs, and CVNs. The shipyards are also home to regional repair centers, which provide planning yard functions, intermediate-level maintenance on equipment, maintenance of key national security infrastructure, and systems maintenance and modernization for special projects.[1]

---

[1] Key infrastructure vital to national security is embedded and maintained at the four Navy public shipyards. These facilities include the only government-owned dry docks capable of docking a nuclear aircraft carrier and certified for docking nuclear carriers and sub-

The NSY mission has evolved and expanded in the past decade. The shipyards are now responsible for managing and executing with broad regional maintenance responsibilities.[2] This means that the shipyards are now responsible for not only the work occurring within their gates but also any maintenance work occurring within the same region at other privately owned shipyards. They provide management and oversight of work that is contracted out to the private sector.[3] This burden generates increased manpower demand.

In addition to the public shipyards, the Navy also performs voyage repairs at sites around the world. These include Bahrain; Groton, Connecticut; Guam; Jacksonville, Florida; and Yokosuka and Sasebo, Japan. Currently, the four public depots provide a significant amount of specialized skills and manpower to these sites. Work performed at these sites is limited by current government regulations and site-specific infrastructure.

---

marines. Additionally, the NSYs have deep-water berths, piers, and wharfs for Navy ships and submarines and large gantry and portal cranes certified for nuclear maintenance. The four shipyards also contain unique machine-shop plant equipment and facilities required for maintenance of the Navy's capital vessels.

[2] Two of the shipyards—PSNS & IMF and PHNS & IMF—have integrated the regional maintenance activities. At Puget Sound, the intermediate maintenance facility at Naval Submarine Base Bangor and intermediate maintenance activity at Naval Station Everett were integrated fully into the naval shipyard in 2002. The Bangor facility performed maintenance and modernization on *Ohio*-class SSBNs, and the Everett activity performed I-level maintenance on homeported surface ships at Naval Station Everett. The Puget Sound and Pearl Harbor shipyards now include a larger, fully integrated regional fleet maintenance and modernization execution and oversight role, in addition to oversight and contracting of private-sector work within shipyard-led availabilities.

[3] The Navy's public shipyards are designated as lead maintenance activities for the fleet maintenance availabilities that they plan and perform. As such, the shipyards are responsible to the fleet, via the type commander and NAVSEA, for final certification of work completion for all maintenance performed. This includes private-sector work performed in these availabilities, which requires the Navy shipyards to integrate all work into an executable and safe overall plan and to maintain oversight of work process controls. Additionally, the two shipyards that have integrated regional intermediate maintenance facilities into the shipyard (Puget Sound and Pearl Harbor) have a contracting role in overseeing work performed by the private sector under multi-ship, multi-option (MSMO) contracts, and other contracting vehicles.

Private Shipyards

In 2015, the Maritime Administration reported: "Currently there are 124 shipyards in the United States, spread across 26 states[,] that are classified as active shipbuilders. In addition, there are more than 200 shipyards engaged in ship repairs or capable of building ships but not actively engaged in shipbuilding."[4] The industry employs nearly 110,000 people.

Of those shipyards, only a few meet the standards to support the Navy. The Navy is currently going through the process to update the list of providers that have an Agreement for Boat Repair (ABR) or a Master Ship Repair Agreement (MSRA). The MSRA is given to shipyards that the Navy has identified as having "technical and facilities characteristics which a ship/boat repair contractor must possess and maintain to ensure that the repair effort on a naval vessel is accomplished satisfactorily."[5] The ABR is given to companies that "demonstrate managerial capabilities to schedule and to control boat/craft repairs. . . . Specifically, an ABR holder must have the management, production, organization and facilities to accomplish the scope of work defined by MSRA."[6]

In lieu of a master list of MSRA- and ABR-certified shipyards, the study team pursued other means to identify private shipyards. The study team searched contract archives for ship repair contracts.[7] The team then compared the archives with what is advertised on company websites (some identify their certifications). The Navy Data Environment, which records data on the completion of availabilities, was used to cross-check the list of providers identified in contracts. Finally,

---

[4] Maritime Administration, *The Economic Importance of the U.S. Shipbuilding and Repairing Industry*, Washington, D.C., November 2015.

[5] Commander, Navy Regional Maintenance Center (CNRMC), *Master Agreement for Repair and Alteration of Vessels; Master Ship Repair Agreement (MSRA) and Agreement for Boat Repair (ABR)*, CNRMC Instruction 4280.1, Norfolk, Va.: Department of the Navy, July 2, 2015, p. 3.

[6] CNRMC, 2015, p. 4.

[7] Searchable archives are available at www.defense.gov (pulled on October 15, 2016).

the study team compared the list of providers with lists developed by others, such as the Government Accountability Office.[8]

The study team identified nearly two dozen private shipyards across the United States, shown in Figure 3.2. General Dynamics owns and operates four shipyards that support Navy ships. These shipyards are located in Bremerton, Washington; San Diego, California; Norfolk, Virginia; and Jacksonville, Florida. BAE also owns and operates four shipyards that support Navy ships. These shipyards are located in Honolulu, Hawaii; San Diego, California; Jacksonville, Florida; and Newport News, Virginia. BAE is the only private shipyard in Hawaii. Two shipyards are owned and operated by Pacific Ship Repair and Fabrication; one is in Bremerton, Washington, and the other is in San Diego, California. In the Pacific Northwest, there is also the Vigor

Figure 3.2
Shipyards Supporting U.S. Navy Warships

Bremerton, WA
- GD NASSCO*
- Vigor*
- Pacific Ship Repair and Fabrication*

Pearl Harbor, HI
- BAE

San Diego, CA
- BAE
- Continental Maritime –HII*
- GD NASSCO*
- Pacific Ship Repair and Fabrication*

Sturgeon, WI
- Bay Shipbuilding

Mobile/Bayou La Batre, AL
- AustalUSA

Pascagoula, MS
- Ingalls Shipbuilding

Westwego, LA
- Avondale Shipbuilding

Kittery, MA/ Groton, CT
- GD Electric Boat
- GD Bath Iron Works

Chesapeake/Norfolk/Newport News, VA
- BAE
- GD NASSCO*
- Marine Hydraulics International*
- Newport News Shipbuilding–HII
- Colonna's Shipyard

Charleston, SC
- Detyens

Jacksonville, FL
- BAE
- GD NASSCO*

SOURCE: Maritime Administration, 2015.
NOTES: * = known Master Ship Repair certification. GD NASSCO = General Dynamics NASSCO; HII = Huntington Ingalls Industries.
RAND RR1951-3.2

---

[8] U.S. Government Accountability Office, *Military Readiness: Progress and Challenges in Implementing the Navy's Optimized Fleet Response Plan*, Washington, D.C., GAO-16-466R, May 2016.

shipyard in Seattle, within Vigor Industrial. San Diego also has Continental Maritime, a subsidiary of Huntington Ingalls Industries.

Industrial Base Assessment

Labor

To assess the industrial base, the study team employed a variety of methods. The study team met with industry representatives from nearly every company shown in Figure 3.2. During these meetings, information about the workforce, facilities, and capacity of the shipyards was collected. Additional information on capacity and resource requirements was provided by the CNRMC. Open-source documentation on the capabilities of the industrial base—including estimates of personnel employed, labor market forecasts, and financial data—were used to supplement information received by industry and the Navy. The following paragraphs summarize the data and information we were able to collect and our interpretation of that information and data.

The public shipyards, which employ nearly 35,500 people, are the dominant provider of U.S. Navy warship maintenance. In the regions where there is a public shipyard that supports surface ship maintenance, the public shipyard employs nearly four times as many people as the sum of private-sector counterparts.[9] Table 3.1 shows the estimated employment levels by region for 2016, in both the public and private sectors. The estimates of the private sector do not include subcontractors or those shipyards with a focus on shipbuilding (i.e., HII-NNS). While subcontractor data would provide a more holistic view of overall capacity, the trade associations and providers did not possess this data, noting that the market has generally responded to their needs, with no major production shortfalls or schedule delays attributable to lack of contractor support. As a result of the missing data, the estimates should be taken as an absolute minimum.

---

[9] This excludes the subcontractors who support the shipyards and excludes HII-NN and Electric Boat, as well as shipyards that solely produce ships.

Table 3.1
Civilian End Strength and Estimated Employment Levels at Public and Private Shipyards, by Region, in 2016

Region	Public Sector:[a] Civilian End Strength	Private Sector:[b] Estimated Employment
Northwest	12,340	2,930
Southwest	0	5,400
Pacific	5,079	750
Northeast	5,510	0
Mid-Atlantic	10,642	2,675
Southeast and Gulf Coast	0	1,250
Total	33,571	13,005

SOURCE: Maritime Administration, 2015.
[a] This includes only public shipyards and excludes RMCs.
[b] This excludes HII-NN, General Dynamics–Electric Boat Bath Iron Works, Austal, Avondale, HII–Ingalls, and Bay Shipbuilding.

Ship maintenance and repair, as reflected in the demand analysis, remains and will remain heavily oriented toward the repair and maintenance of equipment and machinery that, while technically complex, still requires skills more associated with an industrial rather than a knowledge-based economy. Such trends as robotics or autonomous systems may affect some of the way the labor is provided, but the physical realities of where ships and submarines operate, and how they are maintained, will not change. Providers of naval maintenance have experienced shortages in the provision of maintenance services. For example, labor shortfalls have already begun to affect the completion of public shipyard availabilities for submarines, sometimes as long as doubling the availabilities. This does not, in general, seem to be due to lack of facilities or lack of funding. Our interviewees reported that lack of qualified personnel was the most important reason rather than lack of facilities or funding. The resulting delays then trigger the deferral impacts discussed earlier. The public shipyards draw their employees from the same pool as the private providers; our interviews with private provider management indicated considerable challenges providing for some types of labor skills required—in particular, specialized welders and experienced project managers.

Bureau of Labor Statistics projections (Table 3.2) indicate that the national demand for ship repair–related trades is expected to increase

Table 3.2
Bureau of Labor Statistics Outlook on Ship Repair Trades, 2014–2024

Trades	Expected Annual Growth Rate in Demand	Growth Compared to Expected Growth in the Overall Economy (6.5%)		
		Slower Than Average	Average	Faster Than Average
Marine engineers and naval architects	9%			X
Sheet metal workers	7%		X	
Welders, cutters, solderers, and brazers	4%	X		
Metal and plastic machine workers	–13%	X		
Assemblers and fabricators	0	X		
Plumbers, pipefitters, and steamfitters	12%			X
Painting and coating specialists	0	X		
Industrial machinery mechanics	16%			X
Machinist and tool and die makers	6%		X	
Heating, ventilation, air conditioning, and refrigeration technicians	14%			X
Electrical and electronics installers and repairers	4%	X		
Carpenters	6%		X	
Boilermakers	9%			X

SOURCE: Bureau of Labor Statistics, "Employment Projections," web page, undated.

Table 3.3
Man-Days Required per Year in the Port to Maintain the Private-Sector Capabilities-Estimated by Ship Repair Associations

Region	Desired Man-Days	Minimum Man-Days	Maximum Man-Days	FY17 to FY19 Man-Days Expected
Northwest	340K–380K	201K	502K	255K–365K
Southwest	NP	NP	NP	1.3M–1.6M
Hawaii	NP	NP	NP	72K–200K
Mid-Atlantic	1.5M–1.6 M	1.25M	2.0M	1.1M–1.5M
Southeast and Gulf Coast	250K	201K	301K	200K–270K

SOURCE: The ship repair associations provided the data to RAND.
NOTE: NP = not provided.

but, except in a few areas, at a rate equal or slower as that of the broad economy. While in the short term this might conceivably enhance the skilled labor available for ship-specific trades, the longer-term impact is likely to be different. Specifically, Navy ship maintenance is conducted in a limited number of geographic areas, with most others not exposed to the kinds of trades and skills employed in ship repair. With prospects for industrial work in other areas uncertain, the number of entrants into these fields across the national economy may be suppressed. Navy ship repair and maintenance may be one of the few places where demand is growing, even as the national supply declines in response to market forces. These patterns raise questions about whether ship repair and maintenance providers will be able to attract sufficient numbers of qualified trades workers in the future.

Workload

We met with representatives from each of the regional ship repair associations, which included the San Diego Ship Repair Association, the Virginia Ship Repair Association, the Puget Sound Ship Repair Association, and the Jacksonville Ship Repair Association. From each ship repair association, we requested an estimate of the level of work the port would need to prevent layoffs. We also asked for an estimate of a desired level of work and a maximum capacity. We then compared these estimates with the Navy's predicted demand for capacity. The results of these requests are shown in Table 3.3.

The San Diego Ship Repair Association did not provide a response to the request for information, noting that capacity is a function of demand. No data were received from the Ship Repair Association of Hawaii. The others provided estimates based on historical data—on average, how much work have they executed, what was the maximum they executed, and so on. The estimates included total work needed for the port, irrespective of the customer, which could be commercial, the Military Sealift Command (MSC), the Coast Guard, the Navy, or anyone else seeking repairs.

The work the Navy expects to provide to the port is above the minimum threshold identified by the ship repair associations, in most years. The Navy's maximum expected workload is between 20 percent and nearly 200 percent of the ports' identified minimums. The range of workload that can be supported by the ports is larger. This indicates that, on average, the Navy alone is providing enough work to the ports to sustain the shipyards. While, historically, the private sector has been able to increase and decrease the workforce as a function of the demand, there is a cost to this behavior, which has not been quantified.

Facilities

Across the public and private shipyards, there are 54 Navy-certified dry docks. Table 3.4 shows the number of docks, by region, that can accommodate each ship class. Of those, only three can accommodate an aircraft carrier. The mid-Atlantic and Northwest have the preponderance of capability in number of dry docks and the number of classes of ship that can be accommodated in the port. However, the number of docks that would actually be used to accommodate the nonnuclear surface ships, shown above, is an overestimation of capacity, because a nuclear boat or ship will always take priority. The dry-dock capacity issues will be explored further in Chapter Five.

Table 3.4
Number of Docks That Can Accommodate Each Ship Class, by Region, as of 2016

Region	CG-47	CVN-68	CVN-78	DDG-51	LCS-1	LCS-2	LHA-1	LHD-1	LPD-4	LPD-17	LSD-41/49	MCM-1	PC-1	SSBN-726	SSN-21	SSN-23	SSN-688	SSN-774
Northwest	4	1	1	4	10	5	4	4	8	5	8	10	10	5	5	5	6	5
Southwest	2	0	0	2	6	3	2	2	4	2	3	6	6	0	0	0	1	1
Pacific	2	0	0	3	5	4	3	4	4	4	4	5	5	2	2	2	3	3
Northeast	0	0	0	1	8	1	0	0	2	1	1	8	8	2	3	1	6	6
Mid-Atlantic	6	2	2	5	15	10	4	3	14	10	12	16	17	3	3	3	11	8
Southeast and Gulf Coast	3	0	0	5	7	4	1	1	5	2	5	8	8	1	1	0	1	1

SOURCE: NAVSEA 04CX, *Survey of Drydocks*, briefing, Washington, D.C., July 14, 2014.

NOTE: LHA = amphibious assault ship; LHD = amphibious assault ship; LSD = amphibious landing ship; MCM = mine countermeasures ship; PC = patrol ship.

Potential Vulnerabilities

The Office of the Undersecretary of Defense for Maintenance Industrial Base Policy uses a methodology, called the Fragility and Criticality (FaC) assessment methodology, for assessing the likelihood that a specific capability will be disrupted. The method suggests assessments of the financial viability of the providers, the capability to replace the provider if disrupted, and other requirements for entry into the market. Table 3.5 presents the vulnerability measures. Where the study team was able to collect information, through open-source data or interviews with industry representatives, the measure was assessed.[10]

Table 3.5
FaC Assessment Methodology Example

Question
How much total sales for the facility are from Navy contracts?
How many firms currently participate in this firm's market for this capability?
What is the dependence on foreign sources for this capability?
To what degree is the market for this capability commercial?
To what degree are specialized skills needed and available to integrate, manufacture, or maintain this capability?
To what degree is defense-specific knowledge required to reproduce this capability, an alternative, or the next-generation design?
Are specialized equipment or facilities needed to integrate, manufacture, or maintain this capability?
What is the impact on the Department of Defense in time to restore this capability if it is lost?
To what degree are cost, time, and performance-effective alternatives available to meet Department of Defense needs?

SOURCE: Office of the Deputy Assistant Secretary of Defense for Manufacturing and Industrial Base Policy, 2016.

---

[10] Office of the Deputy Assistant Secretary of Defense for Manufacturing and Industrial Base Policy, *Annual Industrial Capabilities Report to Congress for 2015*, Washington, D.C., September 2016.

For the largest of the private-sector providers supporting surface ships, BAE and GD NASSCO, nearly all sales are from Navy contracts. GD NASSCO contracts have been dominated by construction and repair of amphibious ships, while BAE has focused on supporting destroyers, cruisers, and amphibious ships. Given the legal requirement to conduct maintenance activities that are less than six months duration in the ships' homeports, the number of private-sector firms competing in the market is somewhat limited to the providers in the port, which can have as few as two and at most five. Although there is commercial demand for maritime services from ferries, the fishing industry, cruise ships, tankers, and others, the providers that support the Navy require some special skills, facilities, and certifications. In other words, the Navy typically would not place work at a new commercial provider on short notice. The Navy would first apply an evaluation process to certify the provider and the facilities as an MSR or an ABR.

The providers in the Northwest, such as Vigor Industrial, have a diverse portfolio of work, of which the Navy is currently a relatively small part. In recent years, Vigor has been expanding to meet the demands of the commercial sector. If the Navy expects an increase in demand for warship maintenance—which will require the services of Vigor—the Navy needs to procure those services before the capacity is allocated to others. Otherwise, the Navy may need to seek out-of-port solutions to receive the necessary services, at least in the near term. In addition, if there is not adequate dry-dock capacity and a vendor wants to procure or build additional dry docks, the licensing and other requirements can take years.

Summary

Gross estimates of providers and labor indicate a large and robust maritime industrial base in the United States. However, the number of providers servicing U.S. Navy warships is relatively small, particularly if viewed by number of providers in each port. Of the estimated 110,000 people working in the private-sector shipbuilding and repair industry,[11] fewer than half support Navy warships. The amount of work the Navy expects to provide to each port is in line with port need in almost all years from 2017 to 2019. The current providers appear to have a bench of labor they can draw on when demands increase. There is some evidence to suggest that some providers can increase the workforce by nearly 25 percent from one year to the next to support increases in demand. Conversely, the providers can and have implemented layoffs to respond to decreases in demands. Although the shipyards have historically been able to accommodate fluctuations in work, industry has significant concerns about the ability to do this in the future because the industrial base has shrunk and the fluctuations have grown more pronounced. These and other concerns are discussed in Chapter Four. Finally, the number of large dry docks is limited. Chapter Five evaluates dry dock capacity.

---

[11] Maritime Administration, 2015.

CHAPTER FOUR
Industry Incentives Within the Maintenance Industrial Base

The Navy directly controls what happens in public shipyards in terms of workforce and infrastructure development. However, the Navy relies on private industry to make plans and deliver services for surface ships and a large portion of the budget for aircraft carriers. The Navy cannot compel the delivery of these services; it has to create incentives for industry to deliver them and to make capital and personnel-development investments to meet needs over the long term. The market itself is unique in important ways; economic analysis alone can give only partial insight into the way this market works. Another way to approach the issue is to simply ask the actors what motivates them and what they view as challenges.

To support this understanding, the study team met with the ship repair associations and RMCs in Hampton Roads, Puget Sound, Jacksonville, and San Diego. With the help of the ship repair associations and RMCs, the study team also met with representatives from HII-NNS, CMSD, BAE, GD NASSCO, Vigor Industrial, Pacific Ship Repair, and MHI. From these meetings, several themes emerged that bear complete examination. Our interviews with industry representatives did not necessarily result in a uniform perspective on actions by the government. However, a few consistent themes were repeated throughout the many interviews with businesses and organizations. Representatives did not wish to be identified by name and organization, so the following inputs are intentionally not specific.

First, the Navy's planning horizons for availability execution regularly put providers in a position where thorough planning is difficult, resulting in missed maintenance or expensive additions to work packages; these horizons also make it difficult to plan for longer-term capital investments. Second, the Navy has tried different contracting regimes to encourage competition or other outcomes. The most recent, an attempt at firm fixed-price contracts, shows considerable evidence of being counterproductive in terms of encouraging long-term planning. Finally, there are challenges associated with securing the right infrastructure—dry docks in particular—as well as skilled personnel when needed. The Navy only has limited control over these factors, however, and must find a way to bear greater influence by strategic partnering.

Planning Horizons

Private repair shipyards are for-profit organizations; as such, they must produce profits, on average, over time or risk bankruptcy or other negative consequences. All look toward long-term company health, but they are immediately motivated by a need to secure contracts, receive payments, and plan the work they receive.

The Navy's continued inability to provide sufficient planning time for either short- or long-term objectives was an issue repeatedly voiced by industry management. Industry views Navy planning for maintenance as tactical. The short timelines between the request for proposals and the need to begin the availability (contract award), combined with the uncertainty over the amount of future work, is particularly challenging. The maintenance providers expressed concern regarding the quality of the work-specification package produced by a third-party industry team for the RMC. The late contract award date, relative to the availability start date, in addition to a work-specification package that was likely to have changes to it upon execution, represents significant schedule risk to the Navy.

A primary concern presented by some repair associations and their contractor bases was a lack of consideration for the industrial base and sustainment issues in the ship homeporting-assignment process. An example is the assignment of an amphibious expeditionary group that deploys together into a homeport, representing a major fraction of the maintenance workload in the harbor. This construct presents sharp workload-profile changes, from overload conditions to workload levels below the minimum sustainable without large layoffs. Local contractors have had such cycles and offered their observations that workers let go and not brought back within a few months never return and instead pursue other works paths.

Contracting Mechanisms

Industry representatives expressed significant concern over the new contracting environment. Most ship maintenance availabilities will be competed and awarded under a firm fixed-price contract. Although the MSMO contracting vehicle helped industry to deal with these issues in the past,[1] they are no longer being awarded for surface ships. Under MSMO, a single provider was awarded a five-year contract for the availabilities of a certain class of ship. The provider then subcontracted to others in the port. The long planning horizon and steady stream of work allowed for continuous communication between the Navy and industry with respect to the condition of the ships and what might be needed to maintain them. In addition, the providers were able to collaborate to ensure a level loading of the port.

Industry representatives noted that the cooperation that was prominent under MSMO is not possible in the new contracting environment. If a different provider is selected for each availability, the shared knowledge of the ship's condition might not be transmitted, and with each availability, the provider is identifying as new the conditions that may in fact be of long-standing. As a result of the competition, the suppliers will become more competitive and less collaborative,

---

[1] MSMO contracts were awarded to a single provider for a number of availabilities over a five-year period, which helped the shipyard plan resources and make investments in people and facilities.

which may leave some providers without work. The Navy is a single point of business for many of these shipyards. The labor pool is finite in the fleet concentration areas. If one yard wins, another loses and may be forced to leave the business. The Navy relies on the shipyard teaming arrangements to ensure that the capacity at each port meets demands. That reduction in capacity hurts the Navy in the long term, especially if the fleet size increases. The Navy will also be contracting directly with small businesses, which used to be subcontractors to the prime contractor holding the MSMO. The research discovered that the larger companies had been approached by small businesses to help with this contracting. The small business needed help with the management of the Navy contract, because they did not have the infrastructure or management in place to handle government contracting.

The uncertainty in future workload also precludes some from making investments in facilities and people, which introduces risk for the Navy. Specific examples of investments that were approved under the old MSMO contracting environment but that would not have been approved under the current environment underscored the impact of the challenges of the new contracting environment. This includes major investments, such as new dry docks, facility expansions, and other necessary improvements that the Navy relies on the private sector to make. Interviewees noted that the incentive to modernize older facilities is just not there and, hence, is not being done.

In short, the new contracting environment introduces more uncertainty into the workload and less stable planning horizons. The new contracting methodology may do more harm than good, and the Navy will see the negative effects in reduced readiness for the fleet and lower material condition of the individual vessels. The contractors who previously worked together to ensure a stable industrial base may be driven out of doing work for the Navy, further exacerbating the capacity issues described elsewhere in this report.

Labor Market and Infrastructure Challenges

There were also a number of concerns raised regarding the workforce. The various ship repair associations and companies told us that workers not recalled within approximately 90 days would migrate to other areas or other industries and seldom return. This increases the importance of the local industrial base and limits providers to finding labor within the port. One company noted that, after a recent layoff, it was only able to rehire only about 20 percent of the employees it wanted.

Moreover, while it is relatively easy to find green labor, it is harder to find experienced labor. Project supervisors and management are especially difficult to obtain, as much of their experience comes from working within the company and at the specific facilities. When there is growth in the port, experienced labor tends to come from other local companies. Some of the smaller private-sector providers in the Northwest, for example, noted that they had lost some of their best employees to the public shipyards, which are hiring significant numbers of people. This implies that growth in non-Navy work can affect the port's ability to provide services to the Navy. In such areas as Bremerton and Seattle, Washington, where there is a significant commercial demand for services, the Navy must be active to ensure that capabilities are present when needed. A related, noted concern is the difficulty in maintaining early career professionals with potential to advance in the corporation. Given the uncertain nature of the work and volatility of the industry, many younger "millennial" workers end up leaving the company in search of more-stable opportunities.

Dry-dock availability was also noted as a significant concern in some areas. Some shipyards need to team up to have access to a dry dock. In some cases, the dry dock needs to host more than one ship. Under the new contracting strategy, this sort of teaming might not be possible. The introduction of more numbers of the LCS-class ships into ports where the dry-dock capacity is limited, such as Naval Station Mayport, may be very challenging to execute. Figure 4.1, for example, shows the man-day minimum, optimal (where the ship repair association wants the work level to be), and surge levels in Jacksonville.[2] As the fleet mix has changed and the coast-wide bids for maintenance have

---

[2] The three levels were defined by the Jacksonville Ship Repair Association and not independently analyzed by RAND and are meant to illustrate that each port has different levels of capacity in which the Navy needs its work level to fall within.

Figure 4.1
Estimate of Current Range of Mayport Ship Repair Capacity in Man-Days

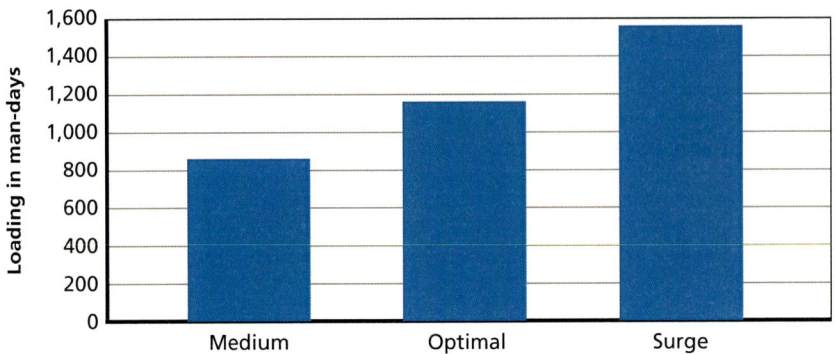

SOURCE: NAVSEA, "WF-300 Workload Allocation and Resource Report (WARR)," spreadsheet, 2014–2016, not available to the general public.
RAND RR1951-4.1

become more commonplace, the man-day level has dropped well below the minimum on several occasions.

The Navy's major fleet concentration ports in the contiguous United States—Norfolk, San Diego, Puget Sound, and Jacksonville—have current capacity profiles that roughly support a range of executable maintenance man-days. Surge capacity can be from resources organic to individual companies or via inflow from either corporate resources or agreements with other companies. Several characteristics of these profiles are important: First, there is a minimum loading whereby companies will, eventually, release trades, engineering, and project resources. Second, the range of margin for surge from desired sustainable levels is on the order of 25 percent, although this varies. Consequently, large changes, such as the assignment of a unique large volume of work, such as cruiser recapitalization, may significantly affect fleet sustainment in the homeport. Similarly, out-of-homeport assignment of an availability during periods of normal or low workload may have a significant, perhaps long-term, impact on the local industrial base for ship repair.

Market Structure

Finally, industry expressed concern that government does not understand the competitive environment that ship repair has become, failing to understand that consolidations have, in fact, decreased the number of competitors, depending on the homeport. Figure 4.2 shows an example of consolidations. This example is the consolidation of multiple competitors into what is now GD NASSCO, which has become one of the two largest providers in San Diego and Norfolk.

Although GD was detailed in Figure 4.2, the same sort of diagram could be produced for BAE, HII-NNS, and a number of other industry providers. There were multiple reasons for these consolidations, but interviews indicated that the reasons center on the expense of facilities, the difficulty that small competitors encounter in not having

Figure 4.2
Consolidation of Ship Repair Providers into GD NASSCO

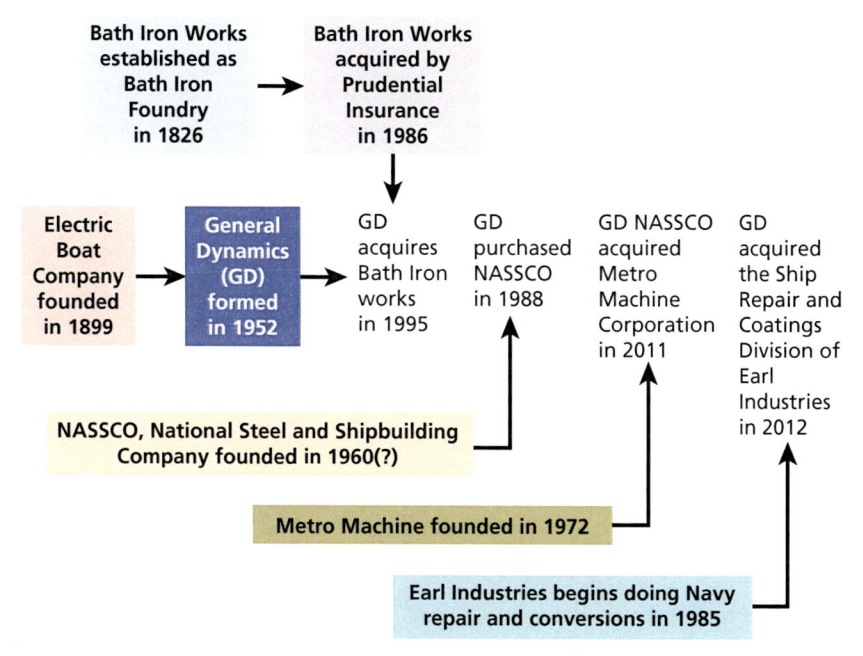

- Bath Iron Works established as Bath Iron Foundry in 1826 → Bath Iron Works acquired by Prudential Insurance in 1986
- Electric Boat Company founded in 1899 → General Dynamics (GD) formed in 1952 → GD acquires Bath Iron works in 1995
- GD purchased NASSCO in 1988
- GD NASSCO acquired Metro Machine Corporation in 2011
- GD acquired the Ship Repair and Coatings Division of Earl Industries in 2012
- NASSCO, National Steel and Shipbuilding Company founded in 1960(?)
- Metro Machine founded in 1972
- Earl Industries begins doing Navy repair and conversions in 1985

SOURCE: GD NASSCO, presentation to the RAND Corporation, September 28, 2016.
RAND RR1951-4.2

a relatively certain demand level, and the inefficiency of several competitors providing identical capabilities. The consequence of this consolidation is that the market becomes less a matter of competition than of planning between one user and a few major providers. The consolidated companies do not fear competition and indeed can respond to it by underbidding competitors possessing fewer resources and flexibility. In fact, the big providers, in general, specialize in particular ship types and work to balance capacity between the providers. When competition is introduced, the main impacts appear to be later award of contracts, artificially suppressed costs, and a loss of some of the benefit that can go with a long-term planning relationship. The market structure does not really provide an environment conducive to competition between a large number of providers that can offer choices to the consumer. The study team further observed, after visits to the four maintenance regions, that, while large national-level companies now represent the majority of the ship repair workforce, these companies do not generally transfer personnel from one homeport to another to meet demand. That is, a subsidiary of a large company located in one homeport that loses a contract to a subsidiary of the same company in another homeport does not transfer a significant amount of workforce to accomplish the work. Industry professionals noted that costs associated with transferring the workforce are significant because of travel, per diem pay, and other reassignment-related expenses. This observation should be considered by the Navy in deciding out-of-homeport availability work assignments.

Summary

The intention of this chapter is not to say that industry is providing a disinterested view of the market that the Navy is obliged to honor. The representatives we interviewed provided a view of a bargaining situation in which they are some of the actors. However, it is also important to understand that these representatives provided a view of the decisions they were likely to make, given a market structure and set of incentives:

- The industry has grown consistently more consolidated, and desire for competition runs against the evolution of the industry into a smaller number of providers. The providers are easily capable of competing against one another, and the actual result of competition is that a small number of providers continue to dominate most markets.
- There are advantages to encouraging a long-term relationship between established providers and the Navy. Knowing the likely long-term demand and requirements allows industry to plan, make risk assessments, and potentially make what could be risky investments in long-term infrastructure and personnel development.
- The personnel skill issue will be challenging and will require a strategic approach by government and industry for workforce development. General economy demand for most repair trades will grow slowly, and there may, as a result, be smaller numbers of qualified personnel seeking to enter the workforce nationally. The majority of workers in ship repair are likely to come from the regions where Navy ship repair is already established. However, this workforce is already losing experienced personnel to age and retirement, and the pace of replacing these people is subject to the time required to build experience. Experienced personnel are, in fact, being drawn from the same labor pool, resulting in cases where the lack of qualified personnel is causing maintenance delays. Recognizing the unique features of ship repair and the workforce to support it, and taking steps to promote greater stability in contract assignment, will be critical to addressing this issue. A strategic approach that does not promote competition among those employers that need skilled workers may be more a more reliable way of securing capability than trying to encourage competition among these employers.

CHAPTER FIVE
Demand-Supply Mismatches

Supply and demand can be compared using two different metrics. The first is labor. Much of the focus when performing supply and demand analysis for shipyards is on shipyard personnel. Demand is that of man-days of maintenance, and supply is personnel capabilities in man-days. The second comparison of supply and demand is of facilities. The supply of facilities can be as much a constraint on maintenance as labor is. Both will be discussed in this chapter.

If plans are executed as written, the resulting private and public demand across time does not show large variations. Annual demand is predictable well into the future, as is the demand for particular kinds of labor and facilities. However, there are several shortfalls that could occur as a result of broader changes in the labor and capital markets.

Labor Market Shortfalls

Long-range workload analysis suggests that the current range of required trade skills will persist, with more emphasis on electrical likely, but there may be challenges in attracting sufficient trade workers in some categories, as discussed in Chapter Three. Navy ships will not change significantly in several important respects that affect sustainment. They will remain large, complicated vessels powered by marine propulsion systems with electrical generation becoming central to the propulsion plants and components and moved by motor-driven shafting and, in some cases, reduction gears directly from the power units (although integrated propulsion systems and electrical drive options may become more prevalent). While there may likely be changes in industrial processes—use of robotics for preservation of confined spaces, for example—there will still be persistent demand for the kinds of skills currently needed to carry out maintenance.

The resulting mismatch does not come from unpredictable demand. It rather comes from the challenges associated with increasing and decreasing the workforce from year to year, in some cases to a large degree, to meet the fluctuating demand. In addition, as discussed, Bureau of Labor Statistics projections show slower than average growth of machinists, fabricators, welders, and electrical workers.[1] Achieving more than entry-level proficiency in these takes years, and demand is sufficient that ship repair facilities may have difficulty finding appropriately trained and experienced personnel. This difficulty is compounded by the competition between public shipyards—that are empowered to hire workers as long-term civil servants—and private shipyards that are affected to a greater degree by fluctuations in demand.

These shortfalls are a matter of particular concern if we consider that labor shortfalls have already begun to affect the completion of public-shipyard availabilities for submarines, sometimes as long as doubling the availabilities. This does not, in general, seem to be due to lack of facilities or lack of funding. Rather, this appears to be a matter of insufficient numbers of qualified personnel. The resulting delays then trigger the deferral impacts discussed earlier. The situation does not right itself quickly.

Other industries have the option of moving offshore or hiring visa holders. Neither appears to be a practical option for this particular industry, where offshore maintenance is specifically proscribed except for voyage repair, and public shipyards are manned by civil servants required to be U.S. citizens. Automation in ship repair may play a greater role—and there are examples from private industry of functions, such as hull cleaning being performed by robotics—but this may

---

[1] Bureau of Labor Statistics, undated.

require some specific investment decisions on the automation of some skills.[2]

Facilities

There has been a significant ramp-up in resourcing the Navy's shipyards with personnel increases over the past four years, and additional, substantial increases are being planned in the near future. There are, however, other factors at the NSYs that may come into play as productivity and capacity constraints. Per interviews with the staff of U.S. Fleet Forces Command, the facilities at public shipyards require recapitalization, and the Commander of Naval Installations Command and NAVSEA have begun examining public-shipyard infrastructure. In the immediate term, public-shipyard infrastructure does not appear to be a major impediment to accomplishing nuclear vessel maintenance. Facility-loading in the private sector has been variable, and there have been cases where port-loading exceeded port capacity. However, the provision of private-sector facilities depends heavily on the incentives offered to industry to provide these facilities. Generally, there do not appear to be delays caused by private-sector shortfalls. However, there is evidence that short-term–focused contracting mechanisms are not giving industry strong incentives for long-term investment. Cranes; machine shop capacity, particularly large capacity outside the machine shop; storage facilities; and pier space are cases in point.

Dry docks present a particular facility investment case worth examining more closely. CMPs call for dry-dock periods whereby more-extensive and more-invasive work is performed, including maintenance to ships' hull, tanks, structural work, more-extensive propulsion plant work and longer-duration modernization ship alterations. Dry docks are specialized facilities, with the number of dry docks of sufficient size and tonnages supported that are certified for US Navy vessels in limited numbers and dispersed geographically. As such, both

---

[2] P. J. Navarro, J. S. Muro, P. M. Alcover, and C. Fernández-Isla, "Sensors Systems for the Automation of Operations in the Ship Repair Industry," *Sensors*, Vol. 13, No. 9, 2013.

public dry docks and private-sector dry docks are key assets to the Navy. Although it is possible for more than one ship to occupy a dry dock, in general, a single Navy ship occupies a single dry dock at a time for a given maintenance availability.

As part of any study on ship maintenance, an examination of how well dry-dock demand matches with dry-dock supply is important. Future dry-dock demand will change given that fleet size may grow and some new ship classes, such as the LCS, require more-frequent dry-docking. To calculate demand, we used CMP-directed dry-docking periods across all ship classes on the East and West Coasts. Dry docks are assigned as available and as directed in the schedules. For the examples below, we assume that the dry docks are filled according to coast-wide bid as opposed to restricted to homeport.

Docking Model Approach

The study team made a number of assumptions to perform the docking facility analysis. First, we used the NAVSEA matrix of approved docks as of 2014,[3] and then augmented this with docks that are expected to come online. To account for different locations, primary missions, and other factors, we added a prioritization scheme. This was based on our assessment of dock capability and resulted in four classifications:

- This dock is the best option and where we most prefer this ship class to be docked.
- The ship class can be docked here, but it is not ideal.
- The ship class can be docked here, but should only be docked here as a last resort.
- The ship class cannot be docked here.

Docks were also assigned to a homeport, fleet area, and a coast. When the analysis was performed by coast, Pearl Harbor was assigned to the West Coast. The one exception is the RCOH dock at HHI-NNS. Because of the unique nature of the dock, the work occurring in the dock, as well as the "heel-to-toe refueling plan" for the carrier fleet,

---

[3] NAVSEA 04CX, 2014.

the dock was separated from the rest. The CMP was used to map out the docking demands on a monthly basis. Once a ship enters dock, it remains there for the duration of the planned docking.

To match ships to docks, the docking plan (based on the CMP) was cycled through the dock-priority scheme to fit the ships to the appropriate docks. Ships that fit in the fewest docks, such as aircraft carriers, were given first priority. If an availability was shorter than six months, we forced the ship to dock within its fleet-concentration area. If an availability was longer than six months, we allowed for coast-wide bidding.

The goal of this scheme is not to assign ships to a drydock with absolute accuracy. The aim was to determine whether the characteristic demand for dry docks matched the characteristic supply. One issue that arises in a 30-year projection is the inability to predict exactly what month ships enter the fleet (for simplicity, January 1 is used for ships not yet in the fleet). Another simplifying assumption is that new ship classes that enter the fleet will fit in the same dock as the classes they are replacing. By the time the model is around 20 years out, phasing problems arise because of the notional ship schedules. This indicates that the model is best used to understand the near-term capacity issues.

Docking Model Findings

Figures 5.1 and 5.2 show that, if the maintenance plans of ships in service or projected are executed, there are multiple years on both the East and West Coasts, where, even with the optimization rules, the model applies—demand exceeds capacity, and ships are displaced.

In both cases examined, there are instances of capacity being insufficient for ship maintenance needs, even in years when overall maintenance demand is lower. This results from suitable dry docks being unavailable for the ships intended to be inducted in a given year. Matching capacity and demand is as much a feat of scheduling as any other consideration. Note that this optimized matching is possible only if the Navy accepts, and utilizes, coast-wide maintenance availability assignment when permitted—with all the crew displacement and challenges associated. Without access to dry docks outside fleet-concentration areas, dry-dock availability is a more significant issue.

Figure 5.1
East Coast Dry-Dock Demand

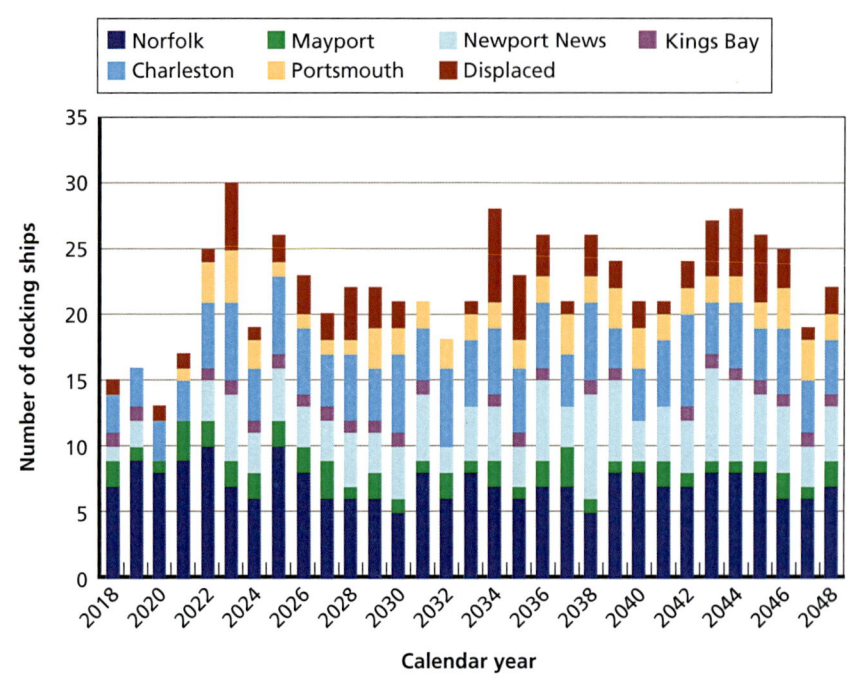

[Stacked bar chart showing number of docking ships by calendar year from 2018 to 2048, with categories: Norfolk, Mayport, Newport News, Kings Bay, Charleston, Portsmouth, Displaced]

SOURCES: Author extrapolation based on CMPs (see Office of the Chief of Naval Operations, 2014) and Office of the Chief of Naval Operations, 2016.
RAND RR1951-5.1

Older docks are creating a supply constraint. As ships get larger, older docks are unable to dock these larger ships. With the inability in many cases to replace old graving docks with new docks, coastal space becomes a new constraint. This will cause the DDG-51, the CG-47, the LCS-2, and other ship classes to overwhelm the supply of docks. Smaller ships are unconstrained in dock space, but with the above classes constituting a large number of the currently active fleet, the model frequently kicks these classes to public shipyards, where the dock problem is a significant issue on the East Coast. After exhausting the few nonsubmarine docks, the model often places these classes in submarine and carrier docks as a last resort. In reality, this is unlikely

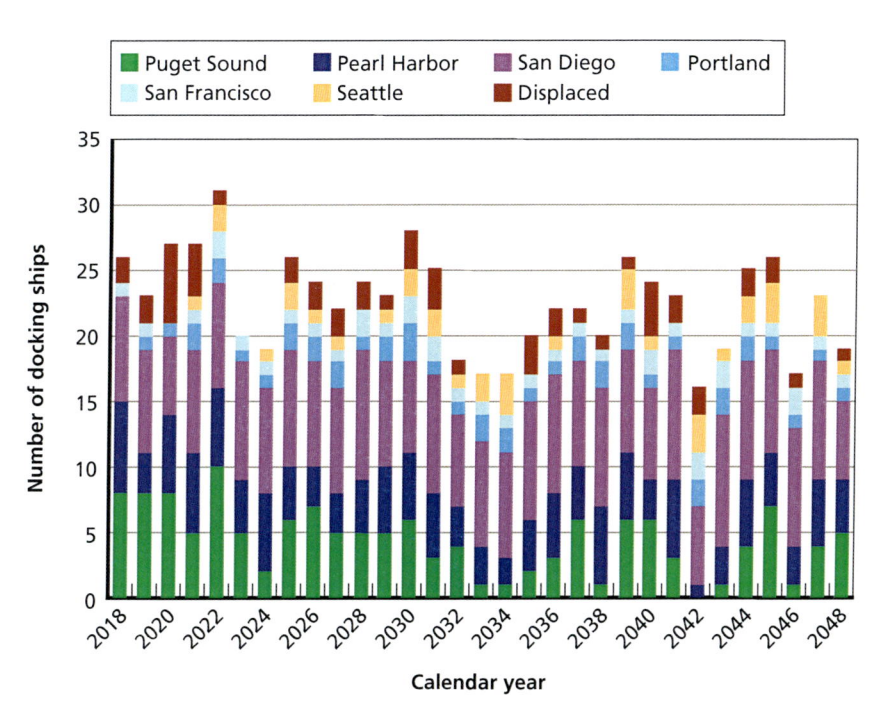

Figure 5.2
West Coast Dry-Dock Demand

[Chart showing number of docking ships by calendar year from 2018 to 2048, with stacked bars representing Puget Sound, Pearl Harbor, San Diego, Portland, San Francisco, Seattle, and Displaced categories.]

SOURCES: Author extrapolation based on CMPs (see Office of the Chief of Naval Operations, 2014) and Office of the Chief of Naval Operations, 2016.
RAND RR1951-5.2

to happen, but it does show a shortfall in the supply of adequately sized docks in the private (and public) docks that are designed for this group.

Dry-Dock Capacity: No Simple Solution

Interviews with industry indicate that the business case for investing in dry-dock capacity is difficult, even with the projected capability shortfalls and capacity issues. Besides being expensive, floating dry docks are not domestically produced, require a variety of environmental clearances, and require some strong assurance that the provider is likely to be the choice for future availabilities. If there is no such assurance, there is little incentive for providers to invest in these projects. Many of the graving docks at the public shipyards are difficult to alter because of shipyard arrangements, historical classification, or both.

Concurrent to the continued build-up of personnel at the NSYs, the Navy would be advised to closely examine other constraints. This is particularly important given that Congress is considering increased force structure, including submarines and aircraft carriers. Any facility investments would need to be well thought out years in advance, and a pathway for such improvements using military construction funding would likely take a decade or more to execute.

Summary

Demand is, to a degree, predictable, although decisions to defer maintenance may make future workload difficult to execute. And supply is, to a degree, subject to the direct control of the government through decisions concerning hiring and investment in public shipyards. The private sector is subject to sets of incentives that can raise or diminish hiring and investment.

There is a mismatch between the trajectory of the future labor force and the needs of the future fleet. This is due to a number of demographic and national economic factors, but the message is that CMPs indicate a demand for certain kinds of specific skills, and the national labor market does not appear ready to provide workers with those skills over time.

There also appears to be a mismatch between key infrastructure supply—in particular, dry docks—and future demand. This will require investment decisions for public-sector shipyards and properly structured incentives for private providers.

CHAPTER SIX

Conclusions and Recommendations

The Navy largely manages demand separately along public- and private-sector providers by platform, even though the supporting funded account, the 1B4B depot maintenance account, is a single account. This approach may need to shift to better use private-sector capacity and expertise to support the NSYs regionally in some areas, such as submarine tank preservation and nonnuclear carrier work in large modernization alterations—for example, to permit the NSYs to focus on core workload that might not be supported externally. This may be an essential focus area if planned fleet force structure commences generating more demand from the shipbuilders that currently augment the ship repair providers and also in the midterm, when additional platforms may be delivered.

If the 30-year shipbuilding plan is executed as planned and if the Navy makes a consistent effort to comply with its CMPs, long-range, future maintenance workload, based on the current Long-Range Shipbuilding Plan for fleet inventory, will remain at least at current levels, with historical trends suggesting that higher maintenance levels are likely. This projection applies in both the public and private sectors. Deferral of maintenance actions will complicate the management of maintenance demands. Deferrals occur for a variety of reasons—including funding shortfalls, scheduling demands, and capacity shortfalls—and it is unrealistic to simply insist that they not occur. However, it is important to understand the impact. Our historical data show that the Navy has shown a tendency to defer maintenance on the two classes of surface ships examined (DDGs and

CGs). Our analysis of the FYDP shows, conversely, that the Navy is planning to spend more than what the technical requirements would have dictated. Our models indicate that this is likely due to an attempt to recover lost maintenance and that the impact on out-year requirements gets more severe the longer the maintenance is deferred. At a minimum, if maintenance is to be deferred, there should be a conscious effort to retire the deferrals on a consistent basis.

The Navy warship maintenance industrial base is characterized by a relatively small number of private-sector providers in each port, two to five, depending on the port. Of the estimated 110,000 people working in the private-sector shipbuilding and repair industry,[1] only a fraction supports Navy warships. The public shipyards, which employ nearly 35,500 people, are the dominant provider of Navy warship maintenance. In the regions where there is a public shipyard that supports surface warship maintenance, the public shipyard employs nearly four times as many people as the sum of its private-sector counterparts.[2] The estimates of the private sector do not include subcontractors or those shipyards with a focus on shipbuilding. This additional information would provide a more complete view of the capacity of the industrial base.

There is evidence to suggest that the current industrial base can meet the expected demands of the Navy. The amount of work that the Navy is planning to provide to the ports between 2017 and 2019 is what the ship repair associations indicate would be required to keep the providers in the port employed, in all ports, in most years. The public shipyards are recovering capacity through the additional hiring of labor, but it will take time for these new hires to become journeymen.

While there appears to be capacity today, there are risks that should be managed to ensure that the capacity is available tomorrow. The private sector has seen significant consolidations in the past 20 years. As a result, there are currently only a handful of privately held companies that perform maintenance on the Navy's surface warships. The

---

[1] Maritime Administration, 2015.

[2] This excludes the subcontractors that support the shipyards and excludes HII-NN and Electric Boat, as well as shipyards that solely produce ships.

Navy needs to continue to provide incentives for companies to stay in the market. Two providers, GD NASSCO and BAE, currently execute the majority of surface warship maintenance. For these larger providers, nearly all revenue is from Navy contracts. GD NASSCO contracts have been dominated by the construction and repair of amphibious ships, while BAE has focused on supporting destroyers, cruisers, and amphibious ships. If one of these suppliers decided to exit the market, the Navy would need time to find alternate providers. Although there are a number of shipyards in the ship repair and construction industrial base, only a few possess the special skills, facilities, and certifications required by the Navy. In other words, the Navy could not place work at a new commercial provider on short notice. The Navy must also provide incentives to industry to make available resources that have not traditionally been used by the Navy. The providers in the Northwest, such as Vigor Industrial, have a diverse portfolio of work, of which the U.S. Navy is currently a relatively small part. In recent years, Vigor has been expanding to meet the demands of the commercial sector. If the Navy expects an increase in demand for warship maintenance, which will require the services of Vigor Industrial, the Navy needs to procure those services before the capacity at these shipyards is allocated to others. Otherwise, the Navy may need to seek out of port solutions to receive the necessary services, at least in the near term.

The type of workload and, hence, the labor skills expected to be required are not likely to change significantly, with a similar distribution by SWLIN of work items appearing consistent in the decades to come. This indicates that trade labor demand by skill will continue to require similar skills to current trades and also new skills to maintain fiber optics systems, photonics, control systems software, and power electronics.

The more significant issue may be the availability of dry docks in homeport. Our analysis of capacity indicated that, if ships are required to remain in port for maintenance, there will be times when docks are unavailable to support the current schedule. Coast-wide bidding can help to alleviate some of the capacity shortfalls, but other approaches will also need to be pursued. Demands for facilities, in particular dry docks, will be significant and, at times, overstress available dry docks by port, but the dry dock demand may be accomplished by allowing coast-wide bidding for dry-docks availabilities. The dry-dock demand predicted for the LCS-1 and the LCS-2, when analyzed by homeport, does not appear executable within available facilities within homeport. Across the public and private shipyards, there are 54 Navy-certified dry docks. Of those, only three can accommodate an aircraft carrier. The mid-Atlantic and Northwest have the preponderance of capability in number of dry docks and the number of classes of ship that can be accommodated in the port, which are driven by the public shipyards.

This report has demonstrated how the Navy's shipyard maintenance industrial base faces challenges in the future. If the most recent 30-year trajectory is correct, demand for a skilled workforce and facilities will grow, and the industrial base, both public and private, cannot support this growth in its current state. Moreover, there are several limitations to maintaining the ability of the Navy to procure the services it needs. For example, broader changes in the labor and capital markets may change how demand is met by the Navy—and the Navy often has little influence over these factors. There are also limitations to how quickly the industrial base can grow before additional constraints and productivity barriers are reached. And there is a cost to sustaining an industrial base that is constantly going through boom and bust cycles.

Despite such deep challenges, however, there are a number of ways to address issues now to lessen future negative impact. We offer the following recommendations for Navy leadership to consider. Together, these recommendations suggest ways for the Navy to move forward by improving overall awareness of the industrial base, better identifying supply constraints, and exploring the impact of changes in procedures.

Work to establish a more integrated picture of port-wide maintenance demands. To improve decisionmaking, it is important for the Navy to develop an integrated picture of public- and private-sector workload that includes commercial, Coast Guard, MSC, and any other maintenance work expected in the port. Where construction yards are relied on to assist with maintenance activities, the construction workload at the private shipyards should also be considered. This will help the Navy identify the availability of resources, prior to execution, which can help to minimize costs. This information can be used to make changes to schedules or to develop the incentives required to secure capacity.

As early as possible in the planning cycle, identify work at public shipyards that is likely to be outsourced. The public shipyards are the largest suppliers of Navy warship maintenance. For aircraft carriers, up to half of the work is outsourced to the private sector. In recent years, because of the capacity constraints at the public shipyards, additional work has been subcontracted. However, the decisions have been made to outsource the work after the work is inducted into the shipyard and determined to be unexecutable, which is more costly to the Navy than contracting directly with the private sector. Identifying the work that is likely to be outsourced on a continuous basis can ensure that the necessary capacity is available and can provide the private sector with a steady stream of work.

Identify expectations for private-sector providers and create incentives for industry to support the plan. Industry would like to have a steady and dependable stream of work, with little fluctuation (up or down). This is the most cost-effective way to operate. It minimizes the costs of training, hiring, and firing, which are a function of the frequency and level of fluctuation. Unfortunately, the work has been characterized by peaks and valleys and other uncertainties. Although the Navy previously provided some stability to industry through a five-year contracting mechanism, the Navy is moving away from this contracting vehicle. While industry has been able to respond to the Navy's demands, historically, industry has expressed significant dissatisfaction with the new contracting environment. It is unclear whether the new environment will provide the incentives required to make investments in people and infrastructure. Communicating expected capacity and capability to the private sector should help.

Explore public-private partnerships as a means to achieve cost and schedule goals. While competition is desired, the number of providers in the space is limited, and without a significant commercial market, competition—or lack thereof—will be determined by the Navy. Although public-private partnerships can be difficult to implement and can exist in many forms, there are significant potential benefits to both the government and industry when implemented well.

For example, identifying and making investments in the facilities and people required to accommodate existing operational and maintenance schedules would become a more cooperative endeavor. The Navy could secure capacity, and industry could obtain more stability. There are many potential challenges with public-private partnerships as well. For example, the government could get locked in to a provider that does not perform or that takes advantage of the relationship by increasing prices. To fully consider the pros and cons of such an approach in the warship maintenance repair community, additional and significant investigation to determine viability is required, but the benefits could be significant, and the Navy should evaluate this approach.

Develop partnered programs for developing ship repairs with specific skill bases. Public and private shipyards compete in the same labor market for the same kinds of skills, some of which are highly specific to ship repair. Moreover, these ship repair facilities are all located in fleet-concentration areas. Exposure to ship repair–related skills are generally not happening at a national level. Per industry interview results, most of the people working in ship repair in Norfolk or Puget Sound, for example, originate from that area. There is little the Navy can do about national trends toward fewer workers available in ship repair–related trades. The Navy can work with industry to develop a career workforce in the areas where ship repair is well established and where there is already at least some base of labor. This will require apprenticeship programs by both public and private maintenance providers, aggressive recruitment at community colleges and among other potential labor sources, and establishment of career paths that allow for the development of a an experienced workforce. The general theme also applies—that future maintenance demands will be better serviced by partnered planning than by competition among a few companies serving a single consumer—applies for the development and management of a career workforce.

APPENDIX A

Shipbuilding and Maintenance Capabilities in the United States, by Region and Shipyard

In this appendix, we present a more detailed look at the capabilities available throughout the United States. We divide the capabilities according to six regions: Northwest, Southwest, Pacific, Northeast and Midwest, mid-Atlantic, and Southeast. We also present relevant details of individual shipyards.

Capabilities in the Northwest

The Northwest region is home to the largest public shipyard, which employs approximately 12,340 people, and supports the widest range of activities—including continuous support to forward-deployed assets. PSNS & IMF, shown in Figure A.1, maintains aircraft carriers in Bremerton, Washington; San Diego, California; and Yokosuka, Japan. Puget Sound can maintain all current and planned aircraft carriers, *Virginia*-class submarines, *Seawolf*-class submarines (currently the sole provider), and surface ships. PSNS & IMF is currently the only public shipyard to perform nuclear defueling tasks prior to decommissioning ships. It also supports SSBNs, whose homeport is in Bangor, Washington, and ships based in Yokosuka.

Unique among the public shipyards, PSNS & IMF supports several off-site locations, including Bremerton, Bangor, and Everett, Washington; San Diego, California; and Yokosuka, Japan. The workload at these sites spans a wide range of platforms and capabilities.

Figure A.1
PSNS & IMF

SOURCE: Provided by PSNS & IMF.
RAND RR1951-A.1

Bangor is the intermediate-level facility for support of *Ohio*-class nuclear ballistic missile submarines. Puget Sound workers at Everett perform continuous maintenance for the USS *Nimitz* (CVN-68) and surface ships stationed with the carrier (with other depot-level work being done at the shipyard). Puget Sound workers use the depot-level facilities in San Diego to perform pier-side maintenance on nuclear ships stationed there. The shipyard usually supplies 600 to 800 workers for six-month planned incremental availabilities for aircraft carriers in San Diego, with nonnuclear work subcontracted to local shipyards. Each year, PSNS & IMF supports the nuclear portion of the annual four-month forward-deployed naval forces ship repair associations performed on the CVN in Yokosuka. The work is similar to the work performed in San Diego. There are currently 47 ships homeported at San Diego (including one Coast Guard cutter), two ships in Coronado, five in Bremerton, 11 at Bangor, and eight in Everett (including two Coast Guard cutter).

In 2016, Puget Sound executed nearly 2.3 million man-days and employed nearly 12,340 civilians.[1] The shipyard has seven dry docks. Only one of the docks can accommodate an aircraft carrier. Five of the

---

[1] NAVSEA, 2014–2016.

docks can accommodate all submarines in the fleet. Four can accommodate all the amphibious ships.

In the Northwest, there is also a robust maritime industry outside the Navy. There are ferries, cruise ships, fishing vessels, tankers, oilers, and others that receive maintenance services in the region. The Coast Guard, MSC, and the Army also pursue maritime vessel construction and or repair services in the region. While not all of these vessels compete for resources with the Navy, some do; there is a maritime ecosystem of people and suppliers that is not easily untangled.

There are three privately held shipyards that provide maintenance services to Navy warships. GD NASSCO, Bremerton, was established in 2014 to support the company's CVN MSMO contract.[2] The site possesses production and planning facilities, as well as piers. The facilities are located near PSNS & IMF and support production, fabrication, and construction of ship equipment for repair and maintenance of CVNs. The company also has trailers located within the public shipyard to provide integrated support to the shipyard. The Navy is the company's primary customer. The shipyard does not have any dry docks. In 2014, the company employed 200 people at this location.[3]

Vigor Industrial offers shipbuilding, ship repair, and industrial services to a wide range of customers at six locations in the northwest. The shipyard has built ferries and catamarans, small and large barges, as well as fishing boats, tugs, cargo vessels and offshore supply vessels. The company has also provided maintenance services to a similar array of vessels, including cruise ships. More recently, the company has started to perform maintenance on Navy destroyers and has installed add-on modules to two of the Navy's Mobile Landing Platforms, but the majority of the revenue of the company is from non-Navy customers.[4] Vigor supports the Navy and Coast Guard in both Portland and

---

[2] General Dynamics NASSCO, "Master Ship Repair from the Pacific to the Atlantic," fact sheet, 2016.

[3] Ed Friedrich, "New Contractor Plans to Hire Locals for Carrier Work," *Kitsap Sun*, October 2, 2014.

[4] RAND interview with Vigor executives, November 3, 2016, Vigor Corporate Offices, Puget Sound, Wash.

Seattle. The company has a NAVSEA-approved dry dock in Portland and two Navy-approved dry docks in Seattle. One of the dry docks is large enough to accommodate an LPD-17; all of the dry docks can accommodate the other amphibious ships. We estimate that the shipyard employs approximately 2,600 workers between the Seattle and Portland locations.[5] This is nearly 600 more people than were employed in 2014.[6] Such growth is an indication of an increasing amount of work.

Pacific Ship Repair and Fabrication has facilities in Everett and Bremerton, Washington. This company also provides an array of services to the Navy, MSC, the Coast Guard, the Army, and commercial companies. It provides ship repair, fabrication, precision cutting, marine closures, and powder coatings.[7]

Each of these companies relies on subcontractors to provide services that are not cost-effective for the shipyards to maintain in house. Discussions with industry representatives indicated that these subcontractors support all of the shipyards as work ebbs and flows across each yard. The shipyard does not have any dry docks. We estimate that the company employs approximately 130 people.[8]

Capabilities in the Southwest

San Diego is a major fleet concentration area with 46 surface ships, three aircraft carriers, and five submarines homeported there. There is no public shipyard in San Diego, but PSNS & IMF supports work in the area, as discussed.

There are four private shipyards that support Navy warships located in San Diego.

---

[5] Linda Baker, "The Love Boat," *Oregon Business*, November 11, 2015.

[6] *Marine Log*, "Vigor Set to Welcome Giant Floating Dry Dock," August 22, 2014b.

[7] Pacific Ship Repair and Fabrication, "Overview," web page, undated.

[8] Buzzfile, "Pacific Ship Repair & Fabrication, Inc.," web page, undated-c; Buzzfile, "Pacific Ship Repair & Fabrication, Inc.," web page, undated-c.

GD NASSCO is the "largest full service shipyard on the west coast."[9] GD NASSCO has been the prime contractor for continuous maintenance activities for several amphibious classes of ships: the *America* class (LHA-6), *Wasp* class (LHD-1), *Whidbey Island* class (LSD-41), and *San Antonio* class (LPD-17). GD NASSCO has also supported the *Oliver Hazard Perry*–class (FFG-7) guided-missile frigates and the LCS classes of ships. The shipyard has also produced a number of ships for the Navy, MSC, and the commercial sector. For the Navy and MSC, the shipyard has produced Mobile Landing Platforms, *Lewis and Clark*–class (T-AKE) dry cargo ship, sealift ships, supply-class ships, hospital ships, prepositioning ships, tenders, oilers, and more. For the private sector, the shipyard has produced tankers, cargo ships, ferries, and other vessels. The shipyard has two certified dry docks. One of the docks can only accommodate smaller ships, such as *Cyclone*-class (PC-1) patrol ships and *Avenger*-class (MCM-1) mine countermeasures ships. The other dock can accommodate all nonnuclear ships. Published reports indicate that GD NASSCO employs nearly 3,500[10] people at this location, but layoffs may occur.[11] Since 2013, the employment level has fluctuated, from 3,200 to 2,800, up to 4,000,[12] and now around 3,500. This suggests that the shipyard has been able to adjust the workforce levels to meet the demands—increasing the workforce by as many as 1,200 people in one year or less.

Continental Maritime is a subsidiary of Huntington Ingalls Industries and repairs ships for the Navy and MSC: "CMSD does repair on all classes of surface vessels including but not limited to CVN, CG, and DDG, as well as all types of amphibious ships includ-

---

[9] General Dynamics NASSCO, 2016.

[10] Chris Jennewein, "NASSCO Warns Employees 700 Layoffs May Be Coming in January," *Times of San Diego*, October 25, 2016.

[11] Brad Graves, "Potential 700 Layoffs at General Dynamics NASSCO," *San Diego Business Journal*, October 25, 2016.

[12] Gary Robbins, "San Diego Shipyards Enjoy Boom Times," *San Diego Union-Tribune*, December 6, 2015b.

ing LHD, LHA, LSD, LPD, and auxiliary ships."[13] The shipyard does not have any Navy-certified docks and employs between 200 and 400 people.[14]

BAE leases its facilities from the Port of San Diego. The company is currently executing five MSMO contracts for the Navy for cruisers, destroyers, amphibious landing ships (LPDs), and mine countermeasures ships. The company has also supported aircraft carriers, large-deck amphibious ships, and frigates. The company also supports cruise ships, tankers, and barges. The company operates two dry docks. One of the docks can accommodate all of the amphibious ships; the other can accommodate a cruiser, a destroyer, and an amphibious landing ship (LSD) or anything smaller. The company employs around 1,600 people at this location.[15] The employment level has fluctuated recently, from 1,450[16] in 2013 to 1,400 in 2014, 2,000 in 2015,[17] and 1,600 in 2016.[18] This indicates that the shipyard has been able to adjust the workforce levels to meet the demands—increasing the workforce by as many as 600 people in one year or less.

Pacific Ship Repair advertises the same services in all locations. These services include ship repair, fabrication, precision cutting, marine closures, and powder coatings. It does not possess a dry dock.[19]

---

[13] Continental Maritime of San Diego, homepage, undated.

[14] Gary Robbins, "San Diego Shipyards to Get $1.3 Billion to Repair Warships," *San Diego Union-Tribune*, March 17, 2016.

[15] California Coastal Commission, *Addendum to Item W13b, Coastal Commission Permit Application #6-15-0555 (BAE Systems San Diego Ship Repair), for the Commission Meeting of May 11, 2016*, San Diego, Calif., May 6, 2016.

[16] Gary Robbins, "Employment Soars at San Diego Shipyards," *San Diego Union-Tribune*, December 24, 2013.

[17] Gary Robbins, "San Diego Shipyards Enjoy Boom Times," *San Diego Union-Tribune*, March 26, 2015a.

[18] Data provided by Southwest Ship Repair Association.

[19] Buzzfile, "Metro Machine Corp.," web page, undated-a.

Capabilities in the Pacific

There are currently two shipyards in Hawaii that provide the majority of maintenance services to the Navy: PHNS & IMF and BAE. These shipyards support the 29 ships that are homeported in Hawaii. BAE's shipyard in Pearl Harbor is colocated with PHNS & IMF and the Navy. BAE currently holds MSMO contracts for cruisers, destroyers, and frigates, which are homeported in Hawaii. In 2013, BAE employed roughly 750 people.[20] The facilities include three piers and one graving dry dock.

PHNS & IMF, shown in Figure A.2, holds a strategic position in the Pacific and provides emergency repairs and other services to fleet assets stationed or deployed in the Pacific. The shipyard primarily supports *Los Angeles*–class (SSN-688) and *Virginia*-class (SSN-774) submarines but also does work on *Arleigh Burke*–class (DDG-51) destroy-

**Figure A.2
PHNS & IMF**

SOURCE: Provided by PHNS & IMF.
RAND RR1951-A.2

---

[20] William Cole, "350 at Pearl May Face Layoff," *Honolulu Star-Advertiser*, February 22, 2013.

ers and *Ticonderoga*-class (CG-47) cruisers. It also has the capability to perform work on any surface ship, the *Ohio*-class (SSBN-726) fleet, and the *Seawolf*-class (SSN-21) submarines. The shipyard has four dry docks; one could accommodate a nuclear aircraft carrier, if required (CVN-75 and older), but it is not capable of supporting carriers on a regular maintenance schedule without upgrades and infrastructure investments. Two docks can service destroyers, amphibious ships, and nuclear submarines. In 2016, PHNS & IMF executed nearly 750,000 man-days of work and employed more than 5,000 civilian employees.[21]

Its location, though strategic, gives these shipyards a limited labor pool to draw from. Costs of labor and material are also higher, posing other challenges.

Capabilities in the Northeast and Midwest

PNSY, shown in Figure A.3, provides depot services for *Los Angeles*–class (SSN-688) and *Virginia*-class (SSN-774) submarines. The shipyard has unique capabilities and the technical expertise required for the repair and maintenance of nuclear submarines and frequently sends skilled personnel to assist with work at other sites. The shipyard also provides off-site support for many nonsubmarine tasks. It is within 160 miles of Groton, Connecticut, homeport to 18 submarines (currently seven in the *Virginia* class and 11 in the *Los Angeles* class).

In 2016, PNSY executed nearly 830,000 man-days of work and employed nearly 5,500 civilians.[22]

General Dynamics Electric Boat is located in Groton, Connecticut. The company focuses on designing and building nuclear submarines, but it has also provided support to naval surface ships and commercial nuclear programs.[23] It has three graving docks and one floating dry dock that are used to construct submarines. Electric Boat also does maintenance, modernization, and life-cycle support of Navy subma-

---

[21] NAVSEA, 2014–2016.

[22] Data provided by PNSY.

[23] General Dynamics Electric Boat, "Our Submarines," web page, undated-b.

**Figure A.3
PNSY**

SOURCE: Provided by PNSY.
RAND RR1951-A.3

rines. The company also has repair facilities in Kings Bay, Georgia, and Bangor, Washington, to support the *Ohio*-class submarines stationed there. There are 14,000 employees at Electric Boat across three locations: the shipyard in Groton, Connecticut; the automated hull-fabrication and outfitting facility in Quonset Point, Road Island; and an engineering building in New London, Connecticut.[24]

Bay Shipbuilding is owned by Fincantieri, an Italian shipbuilding company. The U.S. subsidiary of Fincantieri is known as Fincantieri Marine Group. Marine Group owns three shipyards in Wisconsin, one of them being Bay Shipbuilding. Bay Shipbuilding, however, does not do any government work, but one of the other shipyards, Marinette Marine, builds the LCS-1.

Marinette Marine is located on the Menominee River in Marinette, Wisconsin. The facility is 550,000 square feet, with space for "manufacturing, warehouse and receiving," along with the capacity

---

[24] General Dynamics Electric Boat, "Electric Boat History," web page, undated-a.

to build six littoral combat ships simultaneously. Marinette Marine has computer-controlled manufacturing equipment and heavy-lifting capabilities.[25] In addition to the LCS, Marinette Marine has built *Avenger*-class mine countermeasure vessels and patrol craft.[26] In 2015, when the Navy ordered another LCS to be built by Marinette Marine, it increased its workforce by 200, to a total of 2,000.[27]

Capabilities in the Mid-Atlantic

NNSY, shown in Figure A.4, is the largest public shipyard on the East Coast. It is the only public depot on the East Coast capable of drydocking a *Nimitz*-class (CVN-68) or *Ford*-class (CVN-78) nuclear-powered aircraft carrier.

NNSY currently has the skills and facilities required to support all ship classes. It performs work on aircraft carriers, *Virginia*- and *Los Angeles*–class submarines, large-deck amphibious ships, and surface combatants. It also supports *Ohio*-class SSBNs at Kings Bay, Georgia, which currently seven boats, and runs a foundry and propeller center and a materials test lab.

NNSY is next to one of the major fleet concentration areas on the East Coast, including the roughly 60 ships homeported in Norfolk, Virginia. It is in the same geographic area as HII-NNS, General Dynamics Electric Boat, GD NASSCO, BAE, MHI, and the Mid Atlantic Regional Maintenance Center.[28] Norfolk Ship Repair does significant work on aircraft carriers, frigates, and amphibious ships for the public shipyard. All of these organizations compete for labor but also provide a pool of ready workers from which each can draw.

---

[25] Fincantieri Marinette Marine, "Profile," web page, undated-a.

[26] Fincantieri Marinette Marine, "U.S. Navy Vessels," web page, undated-b.

[27] Rick Barrett, "Navy Orders Another Combat Ship from Marinette Marine," *Journal Sentinel*, April 2, 2015.

[28] In 2008, NNSY and MARMAC merged to form the NNSY & IMF.

Figure A.4
NNSY

SOURCE: Provided by NNSY.
RAND RR1951-A.4

In 2016, NNSY executed 1.5 million man-days of work and employed more than 10,642 civilian employees for the fiscal year.[29]

There are five private shipyards that have a NAVSEA-certified dry dock and support Navy warships. There are numerous smaller shipyards and a robust maritime industry supporting a variety of other maritime vessels.

The Navy is BAE Norfolk Ship Repair's primary customer. It also does some commercial work on cruise ships, tankers, ferries, and cargo ships. Currently, BAE Norfolk is executing five-year MSMO contracts for cruisers, destroyers, and amphibious assault ships. It possesses two NAVSEA-approved dry docks. The employment levels at BAE have declined from approximately 2,500 in 2013[30] to nearly 900 in 2016.[31]

---

[29] Data provided by NNSY.

[30] Bill Bartel, "Ship-Repair Industry Is Threatened with Major Cutbacks," *Virginian-Pilot*, September 8, 2013.

[31] Robert McCabe, "BAE Systems Lays Off 170 Workers in Norfolk," *Virginian-Pilot*, March 30, 2016a.

MHI is one of the smaller ship repair and conversion contractors in Norfolk, with approximately 400 employees.[32] Its services include structural repair, steel plate fabrication, paint and coatings, pipe repair, valve repair, electrical services, and machine work. Historically, MHI has done repair work on destroyers and LPDs in addition to work for MSC, the Maritime Administration, and some commercial ships. BAE Norfolk attempted to acquire MHI in 2012, but the buy fell through for unannounced reasons.[33]

Newport News Shipbuilding is a subsidiary of Huntington Ingalls Industries and is "the nation's sole designer, builder and refueler of nuclear-powered aircraft carriers and one of only two shipyards capable of designing and building nuclear-powered submarines."[34] Although NNS-HII is primarily a ship builder, the company performs the RCOH for the *Nimitz*-class (CVN-68) carriers and provides touch labor to the public shipyards. It possesses seven dry docks, but only one can accommodate aircraft carriers. The company employs nearly 22,000 people,[35] making it the largest employer in the shipbuilding and repair industry.

GD NASSCO–Norfolk is the consolidation of two shipyards: Metro Machine Corporation and Earl Industries. Metro Machine was acquired in 2011,[36] and the Ship Repair Division of Earl Industries was acquired the following year.[37] Both companies, Metro Machine and Earl Industries, had branch locations in Jacksonville, Florida, which now make up GD NASSCO–Mayport. GD NASSCO–Norfolk ser-

---

[32] BAE Systems, "BAE Systems Announces Agreement to Acquire Marine Hydraulics International," press release, November 15, 2012.

[33] MHI Ship Repair and Services, homepage, undated.

[34] Newport News Shipbuilding, "Fact Sheet," Newport News, Va., undated-b.

[35] Newport News Shipbuilding, "About Newport News Shipbuilding," web page, undated-a; Robert McCabe, "75 Laid-Off Workers Called Back at Newport News Shipyard, but More Layoffs Are Possible This Year, *Virginian-Pilot*, April 5, 2016b.

[36] General Dynamics, "General Dynamics Completes Acquisition of Metro Machine Corp.," press release, October 31, 2011.

[37] General Dynamics, "General Dynamics to Acquire Earl Industries' Ship Repair Division," press release, June 29, 2012.

vices include engineering and technical support; shipboard hull and tank preservation; structural and pipe-fitting services; mechanical services; coatings; and advanced preservation solutions, such as nonskid coatings and ultra–high-pressure water. GD NASSCO–Norfolk primarily does repair work on amphibious ships, including the LPD-17.[38] It has one NAVSEA-approved dry dock that can accommodate nearly all nonnuclear ships. It employed nearly 800 people in 2015 but indicated the possibility of layoffs if work does not pick up.[39]

Colonna's Shipyard is a family-run company with a wide customer base servicing Navy and Coast Guard ships, state ferries, and commercial tugboats and barges.[40] Colonna's does welding, ship fitting, steel fabrication, and machining and provides pier-side or voyage repairs. It has one dry dock that can service patrol ships, mine countermeasure ships, frigates, and the LPD-4. The company employs nearly 575 people,[41] 25 more than it did in 2015.[42]

Capabilities in the Southeast and Midwest

The capabilities in the Southeast and Midwest are not nearly as centralized as the other fleet concentration areas. There are shipyards that support the Navy located in Mayport and Jacksonville, Florida, Mobile, Alabama, Westwego, Louisiana, and Pascagoula, Mississippi.

Mayport, Florida, is the homeport for an amphibious ready group, three destroyers, and two cruisers. Current plans call for all of the littoral combat ships to also be homeported in Mayport. There are two

---

[38] General Dynamics NASSCO–Norfolk, "Services," web page, undated.

[39] Allison Mechanic, "Layoffs Coming to General Dynamics NASSCO," WTKR, November 7, 2015; Robert McCabe, "Layoffs Planned at Another Norfolk Shipyard as Navy Work Slows," *Virginian-Pilot*, November 7, 2015b.

[40] Robert McCabe, "Other Shipyards Are Laying Workers Off; Colonna's Is Hiring," *Virginian-Pilot*, October 10, 2015a.

[41] *Marine Log*, "Colonna's Shipyard to Add Larger Dry Dock," December 18, 2015.

[42] McCabe, 2015a.

privately owned shipyards, BAE and GD NASCCO, which currently provide maintenance services to the Navy in Mayport.

BAE has a shipyard in Jacksonville, Florida, that primarily does commercial work but with an adjacent location inside the Naval Station Mayport that does repair and modernization work for the Navy. The Mayport facility is currently executing a single five-year MSMO contract for cruisers and destroyers.[43] Employment levels for BAE at this location have ranged from 650 to 800 between 2013 and 2016. It has two NAVSEA-certified dry docks; one can accommodate cruisers and destroyers, and the other can support smaller ships.

GD NASSCO Mayport operates a small facility in Jacksonville, with fewer than 100 employees.[44] The GD NASSCO facilities in Mayport were designed to support Navy ships. In 2014, GD NASSCO–Mayport won a contract for PC-14 (patrol ship) maintenance,[45] and in 2015, it won a contract for LPD021 maintenance.[46] As of April 26, 2016, GD NASSCO and BAE Jacksonville will be working on LCS-1 sustainment. According to an article in *Defense News*, "BAE Systems Southeast Shipyards Mayport LLC, Jacksonville, Florida (N00024-16-D-4319) and General Dynamics NASSCO, Mayport, Florida (N00024-16-D-4320) are each being awarded firm-fixed-price, indefinite-delivery/indefinite-quantity, multiple award contracts (MAC) to support sustainment execution efforts for Littoral Combat Ships (LCS) 1 variant class ship."[47]

Located in Charleston, South Carolina, Deytens Shipyards is the former Charleston Naval Shipyard. It was bought by Deytens in 1996, following the closure of the shipyard after a base realignment and closure. Deytens primarily does commercial repair work but has three

---

[43] BAE Systems, "Jacksonville Ship Repair," web page, undated.

[44] Buzzfile, undated-a.

[45] *Marine Log*, "NASSCO Mayport Wins $19.8 million PC-14 Contract," June 19, 2014a.

[46] Military-Industrial Complex, "Metro Machine dba General Dynamics NASSCO Mayport Contract Details: Defense Contract Under the Navy Awarded to Metro Machine dba General Dynamics NASSCO Mayport on 10/9/2015," web page, undated.

[47] U.S. Department of Defense, "Contracts: Army," Release No: CR-166-16, August 29, 2016.

graving docks approved by NAVSEA. Deytens specializes in ultra–high-pressure water-blasting and houses "33 marine and industrial companies."[48] The company does not currently provide services to the Navy but does have certified dry docks. It employs approximately 450 people.[49]

Located in Mobile, Alabama, Austal is the prime builder for the *Independence*-variant LCS. Austal also has maintenance and service contracts for the LCS to provide post-shakedown availability, industrial post-delivery availability, and post-delivery availability services. Austal provides corrective and preventative maintenance, including electrical, structural, mechanical, and fit-out.[50] Additionally, Austal provides engineering services and program management support. In 2015, it was reported that Austal USA had more than 4,000 employees, which was the highest employment number the company had ever had.[51]

Ingalls is a part of Huntington Ingalls Industries and is located in Pascagoula, Mississippi. Ingalls builds DDG-51, LHA-6, and LPD-17 ships. It owns 800 acres of land along the Pascagoula River. Ingalls currently employs 11,500 people and has built many Navy and Coast Guard ships. Specifically, Ingalls built 35 DDG-51 ships and is currently the builder of record for the LHA-6 and sole builder of the LPD-17.[52] It won a contract in December 2016 for life-cycle engineering and support services on the LPD-17 for the first year, which is common for the ships that it builds. Under this contract, services include "post-delivery planning and engineering, systems integration and engineering support, research engineering, material support, fleet modernization program planning, supply chain management, maintenance, and

---

[48] Deytens Shipyards, "Company Profile," web page, undated.

[49] AES Marine Consultants LLC, "Safety Alert: South Carolina Shipyard Company Fined Over $100K," LinkedIn, June 12, 2015.

[50] Austral USA, "Support," web page, undated.

[51] Alison Spann, "Austal Reaches Highest Employment Numbers Yet," WKRG.com, October 25, 2015.

[52] Huntington Ingalls Industries, "LPD Amphibious Transport Docks," web page, undated.

training for certain San Antonio-class shipboard systems."[53] In September 2016, Ingalls was also awarded an ESRA for an *Arleigh Burke*–class destroyer, USS *Ramage* (DDG-61).[54]

---

[53] Huntington Ingalls Industries, "Ingalls Shipbuilding Awarded $51 million Life-Cycle Engineering Contract on U.S. Navy's LPD 17 Program," press release, December 16, 2016b.

[54] Huntington Ingalls Industries, "Photo Release—Huntington Ingalls Industries Selected to Perform Overhaul Work on USS Ramage (DDG 61)," press release, September 14, 2016a.

APPENDIX B
Cleaning VAMOSC Data

The model used to partially assess the deferred maintenance in Chapter Four takes all work reported for the *Arleigh Burke*–class destroyer through FY 2014. The data were pulled from VAMOSC by ship. We used the portion that reported maintenance dollars spent in current year (2014). Because VAMOSC data are reported in dollars and TFP and CMP requirements are reported in man-days, we converted dollars to man-days using the estimate of $500 per man-day.

In later models converting man-days into cost, we used 2015 reported port rates for the homeport the ship was currently assigned to. This assumption, of course, came with the caveat that current ships would remain assigned to current homeports and all maintenance would be done at the homeport.

References

AES Marine Consultants LLC, "Safety Alert: South Carolina Shipyard Company Fined Over $100K," LinkedIn, June 12, 2015. As of July 13, 2017:
https://www.linkedin.com/pulse/
safety-alert-south-carolina-shipyard-company-fined-consultants-llc

Austral USA, "Support," web page, undated. As of July 13, 2017:
http://usa.austal.com/support-backup

BAE Systems, "Jacksonville Ship Repair," web page, undated. As of July 13, 2017:
http://www.baesystems.com/en-us/product/jacksonville-ship-repair

———, "BAE Systems Announces Agreement to Acquire Marine Hydraulics International," press release, November 15, 2012.

Baker, Linda, "The Love Boat," *Oregon Business*, November 11, 2015. As of July 12, 2017:
http://www.oregonbusiness.com/article/archives-2006-2009/item/15780-the-love-boat

Balisle, Phillip, *Final Report: Fleet Review Panel of Surface Ship Readiness*, U.S. Fleet Forces Command and U.S. Pacific Fleet, February 26, 2010.

Barrett, Rick, "Navy Orders Another Combat Ship from Marinette Marine," *Journal Sentinel*, April 2, 2015.

Bartel, Bill, "Ship-Repair Industry Is Threatened with Major Cutbacks," *Virginian-Pilot*, September 8, 2013.

Bureau of Labor Statistics, "Employment Projections," web page, undated. As of July 11, 2017:
https://data.bls.gov/projections/occupationProj

"Burke: $2 Billion Backlog in Surface Ship Maintenance Hard to Dig Out Of," InsideDefense.com, March 22, 2013.

Button, Robert W., Bradley Martin, Jerry M. Sollinger, and Abraham Tidwell, *Assessment of Surface Ship Maintenance Requirements*, Santa Monica, Calif.: RAND Corporation, RR-1155-NAVY, 2015. As of April 3, 2017:
http://www.rand.org/pubs/research_reports/RR1155.html

Buzzfile, "Metro Machine Corp.," web page, undated-a. As of July 12, 2017:
http://www.buzzfile.com/business/General-Dynmics-Nassco-Mayport-904-249-7772

———, "Pacific Ship Repair & Fabrication, Inc.," web page, undated-b. As of July 12, 2017:
http://www.buzzfile.com/business/Pacific-Ship-Repair.And.Fabrication,-Inc.-360-674-2480

———, "Pacific Ship Repair & Fabrication, Inc.," web page, undated-c. As of July 12, 2017:
http://www.buzzfile.com/business/Pacific-Ship-Repair.And.Fabrication,-Inc.-425-409-5060

California Coastal Commission, *Addendum to Item W13b, Coastal Commission Permit Application #6-15-0555 (BAE Systems San Diego Ship Repair), for the Commission Meeting of May 11, 2016*, San Diego, Calif., May 6, 2016. As of July 12, 2017:
https://documents.coastal.ca.gov/reports/2016/5/w13b-5-2016.pdf

CNRMC—*See* Commander, Navy Regional Maintenance Center.

Cole, William, "350 at Pearl May Face Layoff," *Honolulu Star-Advertiser*, February 22, 2013.

Commander, Navy Regional Maintenance Center, *Master Agreement for Repair and Alteration of Vessels; Master Ship Repair Agreement (MSRA) and Agreement for Boat Repair (ABR)*, CNRMC Instruction 4280.1, Norfolk, Va.: Department of the Navy, July 2, 2015.

Continental Maritime of San Diego, homepage, undated. As of July 12, 2017:
http://www.continentalmaritime.com/

Department of the Navy, budget materials for fiscal year 2018, web page, undated. As of August 18, 2017:
http://www.secnav.navy.mil/fmc/fmb/Pages/Fiscal-Year-2018.aspx

———, President's budget estimates, Operations and Maintenance, multiple years, FYs 2000–2018. As of April 10, 2017:
http://www.secnav.navy.mil/fmc/Pages/home.aspx

Deytens Shipyard, "Company Profile," web page, undated. As of July 13, 2017:
http://detyens.com/company-profile/

Eckstein, Megan, "Ingalls Shipbuilding Launches First Ship Since Destroyer Program Restart," *U.S. Naval Institute News*, March 30 2015. As of April 3, 2017:
https://news.usni.org/2015/03/30/ingalls-shipbuilding-launches-first-ship-since-destroyer-program-restart

Fincantieri Marinette Marine, "Profile," web page, undated-a. As of July 13, 2017:
http://marinettemarine.com/profile.html

———, "U.S. Navy Vessels," web page, undated-b. As of July 13, 2017:
http://marinettemarine.com/us_navy.html

Friedrich, Ed, "New Contractor Plans to Hire Locals for Carrier Work," *Kitsap Sun*, October 2, 2014. As of June 26, 2017:
http://archive.kitsapsun.com/news/local/new-contractor-plans-to-hire-locals-for-carrier-work-ep-646482557-355251461.html/

GD NASSCO—See General Dynamics NASSCO.

General Dynamics, "General Dynamics Completes Acquisition of Metro Machine Corp.," press release, October 31, 2011. As of July 13, 2017:
http://www.generaldynamics.com/news/press-releases/2011/10/general-dynamics-completes-acquisition-metro-machine-corp

———, "General Dynamics to Acquire Earl Industries' Ship Repair Division, press release, June 29, 2012. As of July 13, 2017:
http://www.generaldynamics.com/news/press-releases/2012/06/general-dynamics-acquire-earl-industries'-ship-repair-division

General Dynamics Electric Boat, "Electric Boat History," web page, undated-a. As of July 12, 2017:
http://www.gdeb.com/about/history

———, "Our Submarines," web page, undated-b. As of July 12, 2017:
http://www.gdeb.com/about/oursubmarines

General Dynamics NASSCO, "Master Ship Repair from the Pacific to the Atlantic," fact sheet, 2016. As of April 3, 2017:
https://nassco.com/wp-content/uploads/Repair-Fact-Sheet-2016_web.pdf'

———, presentation to the RAND Corporation, September 28, 2016.

General Dynamics NASSCO–Norfolk, "Services," web page, undated. As of July 13, 2017:
http://www.nassconorfolk.com/services

Graves, Brad, "Potential 700 Layoffs at General Dynamics NASSCO," *San Diego Business Journal*, October 25, 2016.

Huntington Ingalls Industries, "LPD Amphibious Transport Docks," web page, undated. As of August 20, 2017:
http://ingalls.huntingtoningalls.com/our-products/lpd/

———, "Photo Release—Huntington Ingalls Industries Selected to Perform Overhaul Work on USS Ramage (DDG 61)," press release, September 14, 2016a. As of July 13, 2017:
http://newsroom.huntingtoningalls.com/releases/photo-release-huntington-ingalls-industries-selectedto-perform-overhaul-work-on-uss-ramage-ddg-61

———, "Ingalls Shipbuilding Awarded $51 Million Life-Cycle Engineering Contract on U.S. Navy's LPD 17 Program," press release, December 16, 2016b. As of July 13, 2017:
http://newsroom.huntingtoningalls.com/releases/ingalls-shipbuilding-awarded-51-million-life-cycle-engineering-contract-on-u-s-navys-lpd-17-program

Jennewein, Chris, "NASSCO Warns Employees 700 Layoffs May Be Coming in January," *Times of San Diego*, October 25, 2016.

Marine Log, "NASSCO Mayport Wins $19.8 Million PC-14 Contract," June 19, 2014a.

———, "Vigor Set to Welcome Giant Floating Dry Dock," August 22, 2014b. As of July 12 2017:
http://www.marinelog.com/index.php?option=com_k2&view=item&id=7607:vigor-set-to-welcome-giant-floating-dry-dock&Itemid=223

———, "Colonna's Shipyard to Add Larger Dry Dock," December 18, 2015. As of August 23, 2017:
http://www.marinelog.com/index.php?option=com_k2&view=item&id=10271:colonnas-shipyard-to-add-larger-dry-dock&Itemid=230

Maritime Administration, "Shipyard Reports," web page, undated. As of August 18, 2017:
https://www.marad.dot.gov/ships-and-shipping/national-maritime-resource-and-education-center/shipyard-reports/

———, *The Economic Importance of the U.S. Shipbuilding and Repairing Industry*, Washington, D.C., November 2015. As of July 11, 2017:
https://www.marad.dot.gov/wp-content/uploads/pdf/MARAD_Econ_Study_Final_Report_2015.pdf

McCabe, Robert, "Other Shipyards Are Laying Workers Off; Colonna's Is Hiring," *Virginian-Pilot*, October 10, 2015a.

———, "Layoffs Planned at Another Norfolk Shipyard as Navy Work Slows," *Virginian-Pilot*, November 7, 2015b.

———, "BAE Systems Lays Off 170 Workers in Norfolk," *Virginian-Pilot*, March 30, 2016a.

———, "75 Laid-Off Workers Called Back at Newport News Shipyard, but More Layoffs Are Possible This Year, *Virginian-Pilot*, April 5, 2016b.

Mechanic, Allison, "Layoffs Coming to General Dynamics NASSCO," WTKR.org, November 7, 2015.

MHI Ship Repair and Services, homepage, undated. As of July 13, 2017:
http://www.mhi-shiprepair.com

Military-Industrial Complex, "Metro Machine dba General Dynamics NASSCO Mayport Contract Details: Defense Contract Under the Navy Awarded to Metro Machine dba General Dynamics NASSCO Mayport on 10/9/2015," web page, undated. As of July 13, 2017:
http://www.militaryindustrialcomplex.com/contract_detail.asp?contract_id=31769

Moran, Bill, Adm., Vice Chief of Naval Operations, testimony before the Senate Armed Services Committee, February 8, 2017. As of August 18, 2017:
http://www.navy.mil/navydata/people/cno/Richardson/Speech/VCNO_SASC_Readiness_Testimony_Oral_Statement.pdf

Naval Sea Systems Command, *Surface Maintenance Engineering Planning Program LPD 17 Class Technical Foundation Paper*, Washington, D.C., May 23, 2012a.

———, *Technical Foundation Paper for DDG 51 Class*, Washington, D.C., 2012b.

———, "WF-300 Workload Allocation and Resource Report (WARR)," spreadsheet, 2014–2016, not available to the general public.

———, *Surface Maintenance Engineering Planning Program Technical Foundation Paper, LCS1*, Washington, D.C., April 6, 2015a.

———, *Surface Maintenance Engineering Planning Program Technical Foundation Paper, LCS2*, Washington, D.C., May 4, 2015b.

———, *Surface Maintenance Engineering Planning Program Class Depot Maintenance Technical Foundation Paper*, Washington, D.C., various years and for different classes.

Naval Sea Systems Command 04CX, *Survey of Drydocks*, briefing, Washington, D.C., July 14, 2014.

Navarro, P. J., J. S. Muro, P. M. Alcover, and C. Fernández-Isla, "Sensors Systems for the Automation of Operations in the Ship Repair Industry," *Sensors*, Vol. 13, No. 9, 2013, pp. 12345–12374.

NAVSEA—*See* Naval Sea Systems Command.

Newport News Shipbuilding, "About Newport News Shipbuilding," web page, undated-a As of July 13, 2017:
http://nns.huntingtoningalls.com/who-we-are/

———, "Fact Sheet," Newport News, Va., undated-b. As of July 13, 2017:
http://nns.huntingtoningalls.com/wp-content/uploads/2016/07/NNS-Fact-Sheet.pdf

Office of the Chief of Naval Operations, *Representative Intervals, Durations, and Repair Man-Days for Depot Level Maintenance Availabilities of US Navy Ships*, OPNAVNOTE 4700, Washington, D.C., 2004–2016.

———, *Report to Congress on the Annual Long-Range Plan for Construction of Naval Vessels for Fiscal Year 2017*, Washington, D.C., July 2016. As of April 3, 2017:
https://news.usni.org/2016/07/12/20627

Office of the Deputy Assistant Secretary of Defense for Manufacturing and Industrial Base Policy, *Annual Industrial Capabilities Report to Congress for 2015*, Washington, D.C., September 2016. As of July 11, 2017:
http://www.acq.osd.mil/mibp/resources/2015%20AIC%20RTC%2010-03-16%20-%20Public%20Unclassified.pdf

OPNAV—*See* Office of the Chief of Naval Operations.

Pacific Ship Repair and Fabrication, "Overview," web page, undated. As of July 12, 2017:
http://www.pacship.com/information.php#top

Riposo, J., Michael E. McMahon, James G. Kallimani, and Daniel Tremblay, *Current and Future Challenges to Resourcing U.S. Navy Public Shipyards*, Santa Monica, Calif.: RAND Corporation, RR-1552-NAVY, 2017. As of June 26, 2017:
https://www.rand.org/pubs/research_reports/RR1552.html

Robbins, Gary, "Employment Soars at San Diego Shipyards," *San Diego Union-Tribune*, December 24, 2013.

———, "San Diego Shipyards Enjoy Boom Times," *San Diego Union-Tribune*, March 26, 2015a.

———, "San Diego Shipyards Enjoy Boom Times," *San Diego Union-Tribune*, December 6, 2015b.

———, "San Diego Shipyards to Get $1.3 Billion to Repair Warships," *San Diego Union-Tribune*, March 17, 2016.

Shipbuilding History, "Public Shipyards," web page, undated. As of December 19, 2016:
http://www.shipbuildinghistory.com/shipyards/public.htm

Spann, Alison, "Austal Reaches Highest Employment Numbers Yet," WKRG.com, October 25, 2015. As of August 20, 2017:
http://wkrg.com/2015/10/25/austal-reaches-highest-employment-numbers-yet/

U.S. Department of Defense, "Contracts: Army," Release No: CR-166-16, August 29, 2016. As of July 13, 2017:
https://www.defense.gov/News/Contracts/Contract-View/Article/929464

U.S. Government Accountability Office, *Military Readiness: Progress and Challenges in Implementing the Navy's Optimized Fleet Response Plan*, Washington, D.C., GAO-16-466R, May 2016.

Yardley, Roland J., John F. Schank, James G. Kallimani, Raj Raman, and Clifford A. Grammich, *A Methodology for Estimating the Effect of Aircraft Carrier Operational Cycles on the Maintenance Industrial Base*, Santa Monica, Calif.: RAND Corporation, TR-480-NAVY, 2007. As of June 26, 2017:
https://www.rand.org/pubs/technical_reports/TR480.html